ACKNOWLEDGMENTS

The SFI Press would not exist without the support of William H. Miller and the Miller Omega Program, Andrew Feldstein and the Feldstein Program on History, Law, and Regulation, and Alana Levinson-Labrosse.

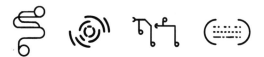

[WORLDS HIDDEN IN PLAIN SIGHT]

*The Evolving Idea of Complexity
at the Santa Fe Institute*

1984–2019

DAVID C. KRAKAUER

editor

© 2019 Santa Fe Institute
All rights reserved.

THE SANTA FE INSTITUTE PRESS

1399 Hyde Park Road
Santa Fe, New Mexico 87501

Worlds Hidden in Plain Sight:
The Evolving Idea of Complexity at the Santa Fe Institute
ISBN (PAPERBACK): 978-1-947864-15-3
Library of Congress Control Number: 2019904092

Chapters 1-22 of this volume were originally published in the *SFI Bulletin*. Chapters 23-36 were first published by Christian Science Monitor as part of the "Complexity" series, supported by Arizona State University. Essays republished with permission, lightly edited for this edition.

The SFI Press is supported by the
Feldstein Program on History, Regulation, & Law,
the Miller Omega Program, and Alana Levinson-LaBrosse.

One has the feeling that this set of samples from the universal Waste Land is on the point of revealing something important to us: a description of the world?

ITALO CALVINO
Collection of Sand (1984)

CONTRIBUTORS

EDITOR & AUTHOR

David C. Krakauer, *Santa Fe Institute*

AUTHORS

Kenneth Arrow, *Stanford University*

W. Brian Arthur, *Santa Fe Institute* and *Palo Alto Research Center*

Rob Axtell, *George Mason University*

Luís M.A. Bettencourt, *Santa Fe Institute*

Seth Blumsack, *Penn State University*

Samuel Bowles, *Santa Fe Institute*

Christa Brelsford, *Santa Fe Institute* and *Arizona State University*

John Casti, *University of Vienna*

Aaron Clauset, *University of Colorado Boulder*

George Cowan, *Santa Fe Institute*

Simon DeDeo, *Indiana University*

Jennifer A. Dunne, *Santa Fe Institute*

Joachim Erber, *Technical University of Berlin*

J. Doyne Farmer, *Institute for New Economic Thinking*

Jessica C. Flack, *University of Wisconsin-Madison* and the *Santa Fe Institute*

Mirta Galešić, *Santa Fe Institute*

Murray Gell-Mann, *Caltech*

David Gray, *Boston Consulting Group*

Marcus J. Hamilton, *Santa Fe Institute*

John H. Holland, *University of Michigan*

NOTE: Affiliations and publication status of works referenced listed in this volume are as of the original essay publication date.

28: The Source Code of Political Power
    Simon DeDeo 281

29: The Complex Economics of Self-Interest
    Samuel Bowles 289

30: Water Management Is a Wicked Problem,
    But Not an Unsolvable One
    Christa Brelsford 295

31: What Can Mother Nature Teach Us About
    Managing Financial Systems?
    Simon Levin and Andrew Lo 303

32: What Happens When the Systems We Rely on Go Haywire?
    John H. Miller 311

33: When an Alliance Comes with Strings Attached
    Paula L.W. Sabloff 317

34: Thanksgiving 2050: To Feed the World We
    Have to Stop Destroying Our Soil
    Molly Jahn 323

35: How Complexity Science Can Help Keep the Lights On
    Seth Blumsack 329

36: Why Predicting the Future Is More Than Just Horseplay
    Daniel B. Larremore and Aaron Clauset 339

37: Emergent Engineering: Reframing the
    Grand Challenge for the 21st Century
    David C. Krakauer 349

C Computation

O Origins & Emergence

M Machines & Culture

P Populations & Collectives

L Life & Evolution

E Economies & Scaling

X eXploration & Theory

HOW TO READ THIS VOLUME

The interconnected nature of complexity science enables you, the reader, to choose your own adventure, as it were. At left are the broad themes tying together the eclectic essays in this book. Follow the shaded tabs on the right edge of the page to pursue a topic.

TIMELINE

THE SANTA FE INSTITUTE ACROSS THE DECADES

SFI ACROSS THE DECADES

The Santa Fe
Institute is founded

Michael Green and
John Schwarz demon-
strate that superstring
theory can only work
in 10 dimensions

Barry Marshall and
Robin Warren show
that ulcers are caused
by bacteria

Fujio Masuoka invents
flash memory

Nintendo
Entertainment system
is released in North
America

David Deutsch's
groundbreaking paper
on universal quantum
computer is published

The Soviet Union
launches the
permanent space
station MIR

Karl Müller and
Johannes Bednorz
discover high-
temperature
superconductivity

Abhay Ashtekar
founds quantum
loop theory

SFI moves into
new digs at the
Cristo Rey Convent

Applied Biosystems
introduces the first
fully automated
sequencing machine

Emerging Syntheses
in Science (David
Pines, ed.) is
published

SFI launches its first
Complex Systems
Summer School

John Miller
becomes SFI's first
postdoctoral fellow

1984

**The Founding
Workshops are held in
October & November**

Attendees include:
Robert M. Adams
Phil Anderson
Herb Anderson
Charlie Bennett
Lewis Branscomb
Felix Browder
Jack Campbell
Peter Carruthers
Stirling Colgate
George Cowan
J.D. Cowan
Irven DeVore
Manfred Eigen
Marc Feldman
Hans Frauenfelder
Harvey Friedman
Murray Gell-Mann
Alan Garen
Mardi Horowitz
B.A. Huberman
Donald Kerr
Edward Knapp
Nicholas Metropolis
Darragh Nagle
David Pines
Ted Puck
Norman Ramsey
Gian-Carlo Rota
John Rubel
M.P. Schützenberger
Doug Schwartz
Alwyn Scott
Marlan Scully
Jerome Singer
Richard Slansky
John Tooby
Anthony Turkevich
Frank Wilczek
Robert Wilson
Stephen Wolfram
Richard Wrangham

1985

Theoretical
Neurophysics

Modeling Evolution,
Evolution of Behavior

Synthesis of
Fundamental Physics,
Astronomy, & Math

Archaeology,
Archaeometry, &
the Extinction of
Flourishing Cultures

Strategies to Model
Troublesome States
of Mind & Associated
Higher Brain Functions

Nonlinear Systems
Dynamics, Pattern
Recognition, & Human
Thought

1986

Adaptive Complex
Systems

Economics: Foraging &
Range Management

International Capital Flows
as a Complex System

Human Genomes &
Peripheral Aspects

Strategies for Avoiding
Major International
Conflict

Heterogeneity &
Public Policy

Project on Fractals

Interdisciplinary
Aspects of Complex
Adaptive Systems

Self-Similarity

Foraging & Hunger

Structure, Dynamics, &
Functions of Proteins

Dynamical Stability in
International Security
Affairs

Mechanisms of
Gene Expression

Sequencing the
Human Genome

Novel Approaches to
Statistical Mechanics

Visualization in
Complex Systems

Modern Approaches
to Large Nonlinear
Physical Systems

Aspects of Self-
Similarity & Scaling

Mathematical Models
in Immunology

Interface between
Computer Science &
DNA Sequencing

Experimental
Evolutionary Novel
DNA-RNA-Proteins

1987

Theoretical
Immunology

The Matrix of
Biological Knowledge

Computational Biology

Evolutionary Paths of
the Global Economy

Climate Change &
Biological Resources

1988

Development of
Theory, Strategy, &
Technology for Applied
Evolution of Polymers
& Biopolymer Systems

Adaptive Networks

Protein Dynamics:
Theory & Experiment

Computer Science &
DNA Sequencing

Elements of
International Stability

Multi-Dimensional
Presentation of Public
Policy Issues

The Economy as an
Evolving Complex
System

Modeling the
Interaction of HIV with
the Immune System

The Elements of Global
Security

The Interface between
Computational Science
& Nucleic Acid
Sequencing

SFI MEETINGS: WORKSHOPS & WORKING GROUPS 1984–2019

Bold text: SFI Milestones *Italic text: World Events*

Christof Koch discovers that at any given moment very large number of neurons oscillate in synchrony and one pattern is amplified into a dominant 40 Hz oscillation

Protests are held in Tienanmen Square in Beijing

SFI kicks off its Research Experience for Undergraduates (REU) program

The Hubble space telescope is launched

The first Internet search engine, "Archie," debuts

CERN publishes first ever web site

William French Anderson performs the first procedure of gene therapy

SFI moves to the Old Pecos Trail Office Compound

Free operating system Linux is introduced

Operation Desert Storm UN Coalition Force led by the US bombs Iraq forces in Kuwait

The first closed-system Biosphere 2 experiment begins

Complexity: The Emerging Science at the Edge of Order and Chaos by Mitch Waldrop is published

SFI Business Network (BNet) is established

The first text (SMS) message is sent from a phone

Gerard 't Hooft develops holographic theory

Charles H. Bennett and others discover how to achieve quantum teleportation using entanglement

1989

Integrative Workshop on the Nature of Adaptive Complex Systems

Adaptation & Learning in Economics

Applied Molecular Evolution/Maturation of the Immune Response

Parallel Computer Systems: Performance Instrumentation & Visualization

Hierarchical Energy Landscapes

Complexity, Entropy, & the Physics of Information, Entropy, & Complexity

Evolution of Human Language

The Economy as an Evolving Complex System

Integrative Workshop on Complex Adaptive Systems

The Organization & Evolution of Prehistoric Southwestern Society

Modeling the Relationship of Human Cognition with Emotion

Learning in Games & Markets

Organization & Complexity in Stochastic Media

US Department of Energy Human Genome Program

Foundations of Developmental Biology

Public Policy

1990

The Organization of Organisms

Glasses, Biomolecules, & Evolution

Artificial Life II

Mathematical Approaches to DNA

Pattern Recognition in Biological Sequences

Organization Evolution Prehistoric Southwest Society

Nonlinear Modeling & Forecasting

1991

Learning in Economics, Psychology, & Computer Science: An Exchange of Ideas

New Technology for Prediction & Pattern Recognition: Applications to Financial Markets

Learning, Rationality, & Games

Third Waddington Meeting on Theoretical Biology

Missing Markets & the Emergence of Market Structure

Price Dynamics & Trading Strategies in Double Auction Markets II

Planning Workshop on Adaptive Computation

Simulation Authoring Tools & SimToolKit Project Symposium

Self-Organized Criticality

Growth & Cities

Implications of Dendritic models for Neural Network Properties

SFI/University of Michigan Adaptive Computation

1992

Theory of Money & Financial Institutions I

Adaptive Processes & Organization

Resource Stress & Response in the Prehistoric Southwest

Founding Workshop in Adaptive Computation

Increasing Returns

Theoretical Computation in the Social Sciences

Biology & Economics: Overlapping Generations

Theory of Money & Financial Institutions II

NATO Advanced Research Workshop on Comparative Time Series Analysis

What OR Models have to Offer CAS & Vice Versa

Artificial Life III

Integrative Workshop: Common Principles of Complex Systems

The Future of Supervised Machine Learning

Audification Workshop

Approaches to Artificial Intelligence

Computation, Dynamical Systems, & Learning

1993

Reinforcement Learning in Robots: The Challenge of Scaling Up

On Learning & Adaptation in Robots & Situated Agents

On Maturational Windows & Cortical Plasticity in Human Development: Is There Reason for an Optimistic View?

Project 2050: Crude Look at the Whole

Scientific Meeting of the SFI External Faculty

Plastic Individuals in Evolving Populations: Models & Algorithms

Theoretical Neurobiology

The Economy as a Complex Adaptive System II

Fluctuations & Order: The New Synthesis

Immune Memory in Theory & Experiment

Founding Workshop on Cultural Evolution

University of Michigan/ SFI Annual Seminar

Artificial World Models

Adaptation & Learning in Organizations

SFI ACROSS THE DECADES

Vice President Al Gore visits SFI for adaptive computation briefing

John Holland gives first annual SFI Memorial Stanislaw Ulam lectures

SFI moves to current home on Hyde Park Road

Nelson Mandela is South Africa's first black president

Jeff Bezos founds Amazon

SFI hosts its first Graduate Workshop

Ted Jacobson derives Einstein's equations of relativistic gravity from purely thermodynamic concepts

The top quark, the last missing quark, is observed at Fermilab

Eric Cornell and Carl Wieman produce the first Bose–Einstein condensate

Ward Cunningham creates WikiWikiWeb, the first "wiki"

Ian Wilmut and his Roslin Institute colleagues clone the first mammal, a sheep named Dolly

The first smartphones are introduced

Giacomo Rizzolatti discovers that the brain uses "mirror" neurons to represent what others are doing

United Kingdom transfers sovereignty over Hong Kong to China

NASA's Pathfinder space probe is the first rover robot on Mars

Toyota begins selling a hybrid car, the Prius

IBM's Deep Blue defeats world chess champion Garry Kasparov in a rematch

Saul Perlmutter, Brian Schmidt & Adam Riess discover that the expansion of the universe is accelerating

James Thomson et al. grow human embryonic stem cells in cell culture

The first handheld "e-readers" are introduced

Jeff Kimble and others teleport a photon for about one meter

George Mitchell uses hydraulic fracturing ("fracking") to extract natural gas from shale

1994

- Complex Systems in Earth Sciences I & II
- Foundations of Genetic Algorithms
- Computational & Molecular Approaches to Evolution
- CLAW (Crude Look at the Whole)
- Searching Sequence Space: Rational & Irrational Approaches to Sequence Design
- Complexity, Entropy, & the Physics of Information III
- Limits to Scientific Knowledge
- Multi-Agent Simulation Systems
- Cold War Science & Technology
- Intervening to Enhance Adaptive Child Development
- University of Michigan/SFI Research Program Seminar
- International Conference on Auditory Display
- Meeting on Project 2050 I & II
- Second HEL Group Meeting on Advances in Genetic Programming
- Issues in Theoretical Neurobiology
- Simulating Organization
- Manufacturing Control

1995

- Women in Science
- Economic ALife
- Graduate Workshop in Computational Economics
- Biological & Computational Landscapes
- Rhythmic Pathologies of the Central Nervous System
- Times to Extinction
- Routines & Other Recurring Action Patterns of Organizations
- The Economy as a Complex Adaptive System
- Sequence Structure Relations in Biopolymers
- Intervening to Enhance Adaptive Child Development
- Ecological Complexity
- ECHO (Ecological & Economic Phenomena)
- Diversity Biotechnology Consortium
- HEL Immune Memory
- Tierra Artificial Life Simulator
- Economics & Finance
- Sloan Centers: Theoretical Neurobiology

1996

- Cellular & Molecular Biology in Health & Environmental Science
- Ecomachines & Spatial Modeling in Ecology & Biology
- Tierra Workshop
- Fundamental Sources of Unpredictability
- Complex Adaptive Systems Approach to Ecological Analysis & Synthesis
- HIV & HIV Structures
- Second Annual Graduate Workshop in Computational Economics
- International Conference on Mathematical Geophysics
- Sloan Centers: Theoretical Neurobiology
- Fundamental Limits to Economic Knowledge
- Mathematical Legacy of R. Bellman
- International Conference for Auditory Display
- Culture Modeling Week
- Dynamics, Computation, & Cognition
- Social Interactions & Aggregate Economic Behavior
- Economics & Cognition Month

1997

- Empirical Analysis of Individual Decision Making
- Universal Phenomena in Ecology?
- ECHO, Sevilleta
- Measuring Relative Rates of Cognitive Development in the First Two Years of Life for Children in Average & Highly Enriched Learning Environments
- Sloan Centers: Theoretical Neurobiology
- Inferential Problems in the Analysis of Treatment Effects
- Interaction-Based Models & the Social Sciences
- Cellular Computation & Decision-Making
- Scaling in Biology
- State & Market Formation
- Limits to Earth Systems Predictability
- Understanding Small-Scale Societies through Agent-Based Modeling
- Evolution of Human Languages
- Sustainability, Inequality, & Growth
- Evolution & Extinction Working Group

1998

- Interplay of Ultimate Causes & Proximate Determinants of Animal Sociality
- Swarmfest '98: Annual Swarm Users Meeting
- Neuro-Mechanical Interactions
- HIV Population Dynamics & Variation
- Energy Flow in Ecological Systems
- Evolution of Scientific Knowledge
- Science at a Crossroads: Roadmapping the Future
- External Faculty Workshop on SFI Integrative Themes
- Adaptive & Computable Economics
- Economics
- Long-Term Human Dynamics
- Towards a Comprehensive Dynamics of Evolution: Exploring the Interplay of Function, Selection, Neutrality, & Accident
- Annual University of Michigan/SFI Workshop
- Constructive Cellular Automata

SFI MEETINGS: WORKSHOPS & WORKING GROUPS 1984–2019

1999	2000	2001	2002	2003
The music downloading service Napster is launched	The first crew is launched to the International Space Station			Grigori Perelman proves the Poincaré conjecture
John Pendry discovers a way to create metamaterials	UC San Diego physicists develop a new class of composite materials with reversed physical properties	Apple introduces the iPod	The Euro enters circulation	The Human Genome Project is completed, having identified all the genes in human DNA
Eric Harris and Dylan Klebold kill 13 people and injure 24 others before committing suicide at Columbine High School, Colorado	Iceberg B-15, with a surface area of 4,200 sq mi, calves from the Ross Ice Shelf of Antarctica	Wikipedia launches Terrorist-hijacked airliner attacks on the US in New York, Washington, DC	Homeland Security Act is signed into law SARS epidemic begins in China	The space shuttle Columbia disintegrates during reentry at the conclusion of the STS-107 mission

1999

- Self-Organization & the Evolution of Development
- Institutions: Complexity & Difficulty
- Embodied Motor Knowledge: Implications for Communication & Learning
- Understanding the Visual Cortex
- Self-Organization & the Evolution of Social Behavior
- Climate & Society on the Colorado Plateau A.D. 500-1600
- SFI Integrative Themes Workshop
- Design Principles for the Immune System & Other Distributed Autonomous Systems
- Empirical Analysis of Social Interactions

2000

- Coevolution of Institutions & Preferences
- Complexity & Agent-Based Modeling
- The Evolutionary Epidemiology of Influenza & Malaria
- Evolvability
- Beyond Equilibrium & Efficiency
- State & Market Formation
- Fellows-at-Large Workshop
- Morphogenesis & Morphological Evolution
- Network Dynamics
- Social Insects: Genes, Neurons, & Societies
- Hierarchies: Distributed Intelligence & the Organization of Diversity
- Mutability: Coevolution of Genes & Diversification Mechanisms
- Computational Approaches to Theoretical Morphology
- Fractals in Biology: Developing the Underlying Mechanistic Principles for Self-Similarity
- Conceptualizing Human Environmental Dynamics

2001

- Evolutionary Dynamics
- Institutional Coevolution
- Structure & Evolution of Strong Reciprocity
- The Internet as a Complex System
- Agent-Based Computing
- New Frontiers for Mathematics
- Hierarchies & Scale
- Poverty Traps
- Disintegrative Themes
- Mathematical Models in Molecular & Cellular Biology
- Computational Complexity & Statistical Physics
- Economic Inequality & Sustainability
- Social Insects: Genes, Neurons, & Societies
- Intergenerational Inequality
- Economy as Evolving Complex System III
- Ecological Complexity & the Sixth Extinction
- Language Emergence & Mathematical Modeling
- Modeling HCV Intra-Cellular Replication
- Early Human Neurobiological & Mental Development
- State & Market Formation
- Network Dynamics
- Behavioral & Evolutionary Economics
- Fellows-at-Large
- Packard Robustness

2002

- Coevolution IV
- Dynamics of Networks & Spatially Extended Systems
- Mathematical Fdns. of Distributed Intelligence
- Robustness/Evolvability of Protein Systems
- Business Complexity: Agent-Based Modeling
- Nonextensive Statistical Mechanics & Thermodynamics
- Biological Framing of Problems in Computing
- Internet Database Analysis/Visualization of Ecological Networks
- Genetics of Phenotypic Robustness
- Globalization & Egalitarian Redistribution
- Network/Supply Chain Optimization
- Social Insect Societies as Multilevel Integrated Systems
- Innovation in Evolution (Prague)
- Biocomplexity for Mathematicians & Physicists
- Evolutionary Innovations Toward an Ecology Based on Principles
- Intervention & Adaptation in Complex Systems (Beijing)
- Defense of Computing Networks
- Generalized Contagion
- Micro-Foundations of Civil War Violence
- Agent-Based Computing

2003

- Linguistic Databases & Taxonomy
- Coevolution of Behaviors & Institutions
- Business Network: Modeling Terrorism in Complex Adaptive Systems
- Robustness of Coupled Natural & Human Systems
- Robust Negotiated Settlements of Civil Conflicts (Bogotá)
- Biophysically Based Functional Models
- Origins & Patterns of Political Violence I
- McDonnell Mini Symposium
- Resiliency & Change in Ecological Systems
- Evolutionary Innovations
- McDonnell Planning Meeting
- SFI Consortium
- Coevolution of States & Markets
- Exploratory Models of Language Acquisition

SFI ACROSS THE DECADES

Google introduces Gmail, which is met with skepticism due to the launch date, April 1

The Harvard-Smithsonian Center for Astrophysics discovers the Universe's largest known diamond, white dwarf star BPM 37093

Andrei Geim and Konstantin Novoselov isolate individual graphene planes

SpaceShipOne becomes the first privately funded spaceplane to achieve spaceflight

The Deep Impact collider hits the comet Tempel 1

YouTube is launched in the US

Rice is the first cereal crop to be sequenced (by the International Rice Genome Sequencing Project)

Hurricane Katrina causes severe damage and kills more than a thousand people on the US Gulf Coast

First BNet Annual Financial Risk Meeting in New York City

The International Astronomical Union relegates Pluto to "dwarf planet" status

Western Union discontinues its telegram service

Paul Rothemund invents DNA origami

French high-speed passenger train, the TGV, reaches 357.2 mph, breaking the record for world's fastest conventional train

Groups led by James Thomson and Shinya Yamanaka discover a way of converting skin cells into embryonic stem cells

Knome introduces the first commercially available human genome sequencing

2004

Origins & Patterns of Political Violence I: Civil Wars

Coevolution of Behaviors & Institutions

Innovation in Natural, Experimental, & Applied Evolution

The Emotional Basis of Cooperation & Punishment

From Structure to Dynamics in Complex Ecological Networks

Prehistoric Chronology: Language, Genes & Migrations

Regions of Innovation

COBRE Center for Evolutionary & Theoretical Immunology (CETI)

Business Complexity: Agent-Based Modeling & Simulation

SFI-Fudan Joint Workshop on Biocomplexity (Shanghai)

SFIC Cognitive Neuroscience

The Dynamics of Groups & Institutions (Ljubljana)

Encoding & Decoding of Biological Signals

Evolution of Human Language: Niger-Kordosanian (Paris)

Scaling Biodiversity (Prague)

Modeling Coupled Natural/Human Systems Over Long Periods

Urban Systems

Analyzing Complex Macrosystems as Dynamic Networks

Biomarkers & Aging

Coevolution of States & Markets

Mathematical & Computer Models in Medicine

Inequality & Social Interactions

2005

Human Social Dynamics

Lying: Methods, Motives, Contexts, & Consequences

Cycles of Social & Environmental Complexity in Lowland Latin America

Encoding & Decoding of Biological Signals

The Road to Software Evolvability

New Perspectives on Complex Systems

Innovation in Natural, Experimental, & Applied Evolution

Adventures in Modeling

Network Robustness to Evolving Agents

Annual Coevolution of Behaviors & Institutions

Innovation in Natural, Experimental, & Artificial Evolution

Language Evolution & Acquisition: Models, Networks, Robustness, & Diversity

States & Markets Planning Meeting

Dynamics of Social Structures

Unifying Current Theories of Ecology

Animal Behavior, Cognition, & Communication: The Role of Networks & Information

Laws of Life

Coevolution of States & Markets

Innovation & Urban Scaling

History & Complexity

A General Overview of Complex Adaptive Systems

Sensitive Data in a Wired World

2006

The Evolution of Gene Regulatory Logic

General Patterns of Migrations

Foundations of Cooperation in the Commons

The Evolution of Inequality: The Long-Term Dynamics of Segmentation, Stratification, & Unequal Reward

From Vent Chemistry to Biochemistry

Degeneracy & Complexity in the Immune System

EARTHTIME III: Probing the Limits of Temporal Resolution in the Geological Record

Modern Malware: Underlying Causes & Potential Solutions

Cognition & Cosmology: New Models for Understanding Mesoamerican Southwestern & Southwestern Relations & Culture Change during the Prehistoric Era

Unifying Current Theories of Ecology

Robustness of Lowland Tropical Rainforests

2007

Dynamic Structure of Robustness

Catalyzing Change in the Global Energy System

Deception: Methods, Motives, Contexts, & Consequences

Complexities of Aging in Biological Systems

Rule-Based Modeling of Biochemical Systems

Nucleic Acid: The First Billion Years

Scaling in Biological & Social Networks

High-Level Perception & Low-Level Vision: Bridging the Semantic Gap

Cosmology & Society in the Ancient Amerindian World

Theoretical Glycobiology: The Search for a Third Language

Models of Emergent Behavior in Complex Adaptive Systems

The Coevolution of Behaviors & Institutions

Ecosystem Models

Inter-Generational Transmission of Wealth in Premodern Societies

The Continued Study of Language Acquisition & Evolution

Language & Genes

Inferring Language History & Prehistory

Southwest Archaeology

Future Challenges in Theoretical Biology

Modern Malware II

SFI MEETINGS: WORKSHOPS & WORKING GROUPS 1984–2019

Barack Obama, the United State's first African American president, is elected

SpaceX Falcon 1 is the first privately developed space launch vehicle to make orbit

London surgeons perform the first successful bionic eye implants on blind patients

CERN's Large Hadron Collider, the world's largest particle accelerator, circulates a proton beam for the first time

Africa's population reaches one billion

Evidence of water is discovered on the Moon

General Motors declares bankruptcy

The H1N1 influenza strain becomes a global pandemic

SFI holds its first Short Course

Craig Venter and Hamilton Smith reprogram a bacterium's DNA

Autonomous vehicles drive 8,000 miles from Italy to China, the first intercontinental trip ever by autonomous vehicles

The Deepwater Horizon oil spill discharges 4.9 million barrels of oil in the Gulf of Mexico

Tohoku earthquake and tsunami in Japan result in nearly 20,000 deaths

Global population reaches seven billion

First successful synthetic organ transplant

Occupy Wall Street protests begin in the US and develop into a global movement spanning 82 countries

Space shuttle fleet is retired

2008

Is There a Physics of Society?
Building Integrative Models of Linguistic Change
Complex Adaptive Systems Thinking in the Study of History
The Role of Variation in Cultural Change
Compartmentation, Phase Separation, & the Origin of Life
Principles of Repurposing
Modeling Technological Innovation
Human Impact of the Last Glacial Maximum
Statistical Inference
New Statistical Approaches to Southwest Archaeology
Life History & Ant–Plant Interactions
From Network Structure to Epidemiological Prediction
The Inheritance of Inequality in Premodern Societies
Social Insects: From Ants to Humans
Reconsidering Counterinsurgency
Dominance, Leveling, & Egalitarianism in Primates & Other Animals
Phylogenetics Workshop
Integrating Evolutionary Theory into Cancer Biology
Principles of Biological Computation
The IHOPE Project
First Steps Toward Understanding Market Ecologies
Networks & Navigation
Complexity & International Relations
Modern Malware III
Cosmology & Society in the Ancient Amerindian World

2009

Collective Decision Making: From Neurons to Societies
Evolution Complexity & the Law
First Annual ITP–SFI Research Workshop on Frontiers in Complex Systems: Complex Social Networks & Urban Dynamics
Systems Biology & the Physical Foundations of Aging
Market Design & Structure
Neuro-Cognitive Niche Construction
Self-Regenerative Approaches to Computer Security
Emergence of Hierarchy & Inequality
Universal Diversity Patterns Across the Sciences
The Complexity of the Gene Concept
The Coevolution of Behaviors & Institutions
Models of Innovation & Propagation in Language Change
From Geochemistry to the Genetic Code
Minimal Life
The Mathematics of Terrorism

2010

The Role of Entropy in Language & Communication
Emergent Properties & Resilience of Interaction Networks
Decentralized Control in Systems of Strategic Actors
Viral Dynamics
Cosmology & Society in the Ancient Amerindian World
Perception & Action—An Interdisciplinary Approach to Cognitive Systems Theory
Peopling of the Americas
The Coevolution of Behaviors & Institutions
Persistent Inequality: The Dynamics of Wealth Inequality in Premodern Societies
Rethinking Aid to Support Economic Growth
Immune Response Consortium
The Ecophylogeny of Complex Species Interactions
Archaeoastronomy Meeting
Reasoning, Perception, & Beliefs in Strategic Settings: Theory, Behavior, & Cognition

2011

Randomness, Structure & Causality
The Coevolution of Behaviors & Institutions
Genomic Imprinting
Frontiers of Data Analysis
Quantum Life
Computational Cultural Evolution
Agent-Based Economic Crisis
Neurobiology Meeting Group
Causality Meeting Group
Practical Methods for Analysis of Early Warning for Regime Shifts

SFI ACROSS THE DECADES

SFI receives $5 million Templeton grant

Floyd Romesberg chemically synthesizes two artificial nucleotides and inserts them into a bacterium, thus creating a new genetic alphabet

PAL-V builds a flying car

Markus Covert simulates an entire living organism (Mycoplasma genitalium) in software

Jennifer Doudna's group invents the CRISPR-Cas9 system for gene editing

SFI publishes its online course website, Complexity Explorer

Art appraiser Eugene Thaw donates 36-acre Tesuque property to SFI.

XBox One is launched

Nearly 1,500 people are injured when a meteor explodes over the Russian city of Chelyabinsk

Benedict XVI becomes the first pope to resign voluntarily in more than seven centuries

Robert Lanza generates human stem cells from adults

Ebola epidemic kills more than 11,000 in West Africa

A 4.4-billion-year-old zircon fragment found in Australia is confirmed as the oldest known piece of Earth's crust

SFI partners with Christian Science Monitor on a multi-author series, "Complexity"

Junjiu Huang genetically modifies human embryos

Gravitational waves are observed for the first time, 100 years after they were discovered by Einstein

Liquid water is discovered on Mars

2012

Dynamics of Inequality in Premodern Societies

Applying Complex Systems Thinking to Program Evaluation

Power Grids as Complex Networks

Can There be a Science of Cities?

The Principles of Complexity: Life, Scale, & Civilization I

Combining Information Theory & Game Theory

The Coevolution of Behaviors & Institutions

Cultural Processes that Give Rise to Social Monogamy

Machine Learning for Materials Design

Legacies of the Manhattan Project: A Case Study in the Consequences of Conflict

Multi-Sampler Optimization

Planning Archaeological Infrastructure for Integrative Science: Grand Challenges in Archaeology

Neurobiology Working Group in Olfaction

Early Maya E-Groups, Solar Calendars, & the Role of Astronomy in the rise of Lowland Urbanism I

Scientist/Artist Research Collaborations

Risk I

Multi-Information Source Optimization

2013

Network Structure, Political Hierarchy, & Inequality

Structure, Statistical Inference, & Dynamics in Networks

The Principles of Complexity II

Deep Computation in Statistical Physics

Technological Evolution & Economic Growth

Rethinking Network Science & Modeling for Critical Infrastructure Protection, Analysis, & Development

How Far Can Big Data Take Us Toward Understanding Cities?

Gateways to Emergent Behavior in Science & Society

Theory & Knowledge Systems for Sustainability Science

Collection, Organization, & Analyses of Community-Based Survey Data

Self-Invention in Biological & Societal Systems

Gradient-Based Ecological Network Research

Machine Learning for Materials Design

Risk II

Social Complexity at Cahokia

Theoretical Biology & HIV

Algebra, Cryptography, & Quantum Algorithms

Big Data in the Brain

Complex Social Science Galaxy Databases & Modeling

Early Maya E-Groups, Solar Calendars, & Astronomy II

From Co-Infection to Cultural Dissonance

Information Theory of Sensorimotor Loops

Similarity & Divergence in Post-Neolithic Europe

2014

Earth League

Statistical Mechanics Foundations of Complexity

Next-Generation Surveillance for the Next Pandemic

Principles of Complexity III

Origins of Novelty

Frontiers in Niche Construction

Acting Locally, Understanding Globally

Dynamics of and on Networks

Conceptual Innovation & Major Transitions in Human Societies

Coevolution of Institutions & Behaviors

Molecular Network Topology & Local Adaptation

Historical Linguistics

Epigenetic Regulation

Agent-Based Modeling of Archaeological Transitions

Cultural Evolution

Social Change in the Context of Climate Challenges

Global Human Ecodynamics Alliance Modeling

Generation, Acquisition, & Diffusion of Knowledge

Settlement Scaling in the Ancient World

Alternative Energy Technologies

Creativity in the Brain

Complex Energy Landscapes

Collective Cognition

Convergent Evolution of Agriculture in Insects & Humans

Network on Inequality, Complexity, & Health

NETI (Chile)

Information Theory, Ecosystems, & Schrödinger's Paradox

The Major Transitions in Artificial Evolution

Universals in Human Biosocial Organization

2015

Dynamics of Wealth Inequality

Reinventing the Grid

Emergent Paradigms in Nonlinear Complexity

Kinetic Networks

Wildness in Computer Science, Physics, & Mathematics

Innovation as Search on a Space of Possibilities

Evolutionary Ecology of Complex Life Investment Strategies

Inference on Networks

Coevolution of Institutions & Behaviors

Dynamic Primate Contact Networks & Disease Risk

Studying Athapaskan

Maya Materialization of Time

Gradient-Based Ecological Network Research II

Matrix Multiplication

Macrohistorical Datasets

Large-Scale Spatial Synchrony in Ecology & Epidemiology

Algebra, Geometry, Pseudo-randomness, & Complexity

Coevolution of Social Structure & Communication

Settlement Scaling in Premodern Societies

Information Engines

Interdependent Networks

Motility in the Immune System

Historical Linguistics Datasets

Human Cooperation

Grassland & Mammalian Community Dynamics

Cell Type Origination

Cybertools for SKOPE

Information Theory, Ecosystems, & Schrödinger's Paradox

Molecular Networks & Evolution

Childhood Obesity Modeling

The Inverse Ising Problem

SFI MEETINGS: WORKSHOPS & WORKING GROUPS 1984–2019

BNet is relaunched as the Applied Complexity Network (ACtioN)

SFI launches website redesign

First SFI Applied Complexity Studio is held

United Kingdom votes to withdraw from the European Union

The Solar Impulse 2 becomes the first solar-powered aircraft to circumnavigate the planet

The augmented reality game Pokémon Go is released, breaking revenue and sales records

SFI Press debuts first volume

25th Anniversary of BNet/ACtioN's longest member, John Deere

Women's March in response to the inauguration of Donald Trump as President is largest single-day protest in US history

Both gravitational and electromagnetic waves from the first observed collision of two neutron stars are detected

The third known species of orangutan is identified in Indonesia

2016

- Re-Emerging Infectious Diseases: The Challenge & Opportunity of Pertussis
- Reinventing the Grid: The Nature of Technological, Social & Industrial Innovation & Transition in Power Generation & Delivery
- Limits to Human Performance
- Limits to Prediction
- Frontiers of Ecological Theory Integration
- Statistical Physics, Information Processing, & Biology
- Studying the Interplay of Hard Modularity & Dynamics: From Lifecycle Strategies to Political Campaigns
- Teaching & Learning Economics as if the Last 30 Years Had Happened
- Comparing Ecological Networks along Gradients
- Centralized vs. Decentralized Control in the Regulation of Populations
- Ecological Data Dramatization for Art & Science
- Combating Sample WEIRDness in the Social & Behavioral Sciences
- Lexical Semantic Networks & Language Change
- Do Population-Level Demographic Processes Accurately Reflect the Joint Contributions of Individuals? Implications for Modeling Human Social Dynamics
- 72 Hours of Science
- Convergent Evolution of Agriculture in Insects & Humans II
- History, Networks, & Evolution
- Stochastic Search Processes

- New Algorithms for Group Isomorphism
- Origins of Large-Scale Spatial Synchrony in Ecology & Epidemiology
- Collective Animal Motion in the Wild
- Toward a More Inclusive Theory of Social Evolution, Integrating Migration & Life History Constraints
- Predicting the Response of Host-Associated Microbiomes to Disturbance
- Social Network Interventions
- Olfaction
- Human Settlements & Networks in History
- The Maya Materialization of Time, Synthesis of Themes
- JSMF–SFI Collaboration on Complex System Science
- Circumventing Turing's Achilles' Heel
- Collective Problem Solving
- Biological Circuit Evolution
- Geometric Complexity Theory

2017

- JSMF–SFI Postdocs in Complexity Conference I & II
- Dynamics of Networks & Inequality
- Harold Morowitz Symposium
- Thermodynamics of Computation in Chemical & Biological Systems
- Thermodynamics & Computation: Toward a New Synthesis
- NSF Ideas Lab: Practical Fully Connected Quantum Computer Challenge
- The Limits of Understanding: Past, Present, & Future
- Strategies in Adaptive Systems: From Life Cycles to Political Campaigns
- Artificial & Natural Intelligence
- Evolution & Restraint of Malicious Behavior in Complex Systems
- Phase Transitions in Human Sociality
- Human-Centered Interaction Networks through Space & Time I & II
- Modeling Dynamics of Violent Radicalization in Western Democracies: Exploring the Feasibility of a Research Collaboration that Connects Micro, Meso, and Macro Levels of Explanation
- Invention & Novelty across Domains: Developing a Research Agenda, Building a Theory
- Origins of Large-Scale Spatial Synchrony in Ecology
- Quantifying Collective Behavior in Living Systems I & II
- 72 Hours of Science II
- Noli Timere: Beneficial Epidemics in a Developing Graphic Novel

- Cooperation & Constructing the Commons in Spatially Complex Communities
- Envisioning New Modes of Cultural & Technological Changes
- The Future of Computational Social Science
- Explaining Significant Household Gini Disparities between the Old and New Worlds in Prehistory
- Telling Time: Myth, History, & Everyday Life in the Ancient Maya World
- Information Networks & the Evolution of Social Organisms
- Morphological Computation
- Agglomeration Economies, Past & Present
- The Nexus of Ecology & Evolution in Space & Time
- Liquid Brains, Solid Brains
- Computational Study of the Law

SFI ACROSS THE DECADES

SFI hosts its inaugural InterPlanetary Festival in the Santa Fe Railyard

Construction begins on SFI's Miller Campus

Apple becomes the world's first trillion-dollar company

The European Union's General Data Protection Regulation (GDPR) goes into effect

A Neanderthal-Denisovan "hybrid" fossil is identified in Denisova Cave, Russia

The northern white rhinoceros becomes functionally extinct when the last male of the subspecies dies in Kenya

The Dow Jones industrial average falls 1,597 points in one day, its largest drop in history

SFI Press publishes second edition of *Emerging Syntheses in Science* (David Pines, ed.)

SFI institutes programming at Miller Campus

Chinese probe Chang'e 4 is the first human-made object to land on the far side of the moon

A second case of sustained HIV in remission is reported, ten years after the first

The Event Horizon Telescope project announces the first ever image of a black hole in the M87 galaxy

Fossil fragments of a new species of human, *Homo luzonensis*, are discovered in the Callao Cave in the Philippines

SpaceIL launches the Beresheet probe in the first privately financed mission to the moon

2018

Integrating Development & Inheritance

JSMF-SFI Postdocs in Complexity Conference III & IV

From Judgment to Impact

Perspectives on Social Learning

Socio-Hydrological Dynamics

Time in Adaptive Systems

Limits to Human Performance

Artificial Intelligence & the "Barrier of Meaning"

Major Transitions in Life: Origins to Translation

Directional Biases in Evolution

The Meaning of Information

Lookahead Optimization in Artificial & Natural Systems

The Network Structure & Dynamics of Cooperation & Competition in Humans

The Complexity of the Patent System

Extending the Reach Infometrics

Coevolution of Institutions & Behaviors

Limits to Inference in Networks & Noisy Data

The Fragility of Growth in the Past

Modeling Hate Crimes

Integrating Perspectives on Social Learning

Oxygen & the Rise of Animals

Universal Features of Decision Making via Collective Computation

Cumulative Cultural Evolution

The Human Niche

Cognitive Regime Shift I: When the Brain Breaks

Aging & Adaptation in Infectious Diseases

The Growing Gap Between our Physical & Social Technologies

Spatio-Temporal Correlations in Ecology

Settlement Scaling & the Challenge of Mayan Low-Density Urbanism

Lexical Semantics, Language Change, & Universal Translators

On Time & Being Maya

Agnostic Biosignatures for Astrobiology

AI and the Barrier of Meaning

Population & the Environment

Human-Centered Interaction Networks through Space & Time III

Managing Natural Risk in the Modern & Prehistoric World

Next-Generation Ecological Network Theory & Application

Dynamic Multi-System Resilience in Human Aging

Theoretical Aspects of the Origin of Life (ASU)

Integrating Critical Phenomena & Multi-Scale Selection in Virus Evolution

Tackling Complex Sustainability Issues

Culturally Transmitted Mate Preferences & the Process of Evolution by Sexual Selection

University Scaling: Mechanisms of Institutional Strategy & Trade-Offs

2019

Postdocs in Complexity Conference V

Self-Organization & Emergence in Nonequilibrium Matter

Collective Crypto

Stuart Kauffman Celebration

What is Biological Computation?

Irreversible Processes in Ecological Evolution

Sociality Under Scarcity

The Role of Information in Complex Conflict

A New Paradigm for Political Economy

Cleaning Up the MESS: A Working Group to Complete the Massive Eco-Evolutionary Synthesis Simulation

Higher-Order Interactions: Experiments, Inference, & Models

Integrating Species Competition, Coexistence, & Network Theory

Cell Type Concept

Complex Time Advisory Board Meeting

The Evolution, Benefits, & Costs Underlying Categorization

Aging in Single-Celled Organisms I & II

Hallmarks of Biological Failure

Culturally Transmitted Mate Preference & the Process of Evolution by Sexual Selection

Thermodynamic & Computational Efficiency in Cellular Chemical Reaction Networks

Aging & Adaptation in Infectious Diseases II

Macroecological Insights into Microbiome Resilience & Function

Integrating Views on Urban Scaling

New Algorithms for Optimization, Sampling, Learning, & Quantum Simulations

Humans in Ecological Networks

What is Sleep?

SFI MEETINGS: WORKSHOPS & WORKING GROUPS 1984–2019

...AND BEYOND

INTRODUCTION

David C. Krakauer, Santa Fe Institute

"They consider only their own ideas of ingenuity; and, in searching for anything hidden, advert only to the modes in which they would have hidden it."

— EDGAR ALLAN POE, "THE PURLOINED LETTER" (1845)

What explains our obsession with the hidden?

Whence comes our enduring belief that truth is not to be found by means of immediate perception but through extreme efforts at augmenting what can be sensed with manufactured instruments? It is as if every scientific project were a crime scene, the perpetrators long since fled, leaving a few clumsy crumbs of evidence for researchers to puzzle over with magnifying lens and graphite powder.

From classical antiquity and its obsession with geometric order beyond manifest reality, followed by the late Renaissance conviction that everything in nature expresses an unfathomable divine intention, on through to the modern age with its instruments required to reveal the elemental nature of reality (microscopes, telescopes, particle colliders, mass spectrometers), what is causally primary is always thought to be buried deep beneath the surface of the obvious.

Traditionally things can be hidden in two fundamentally different ways. Things can be hidden in space, and they can be hidden in time.

To hide in space means that phenomena lie beyond the scope of our everyday senses because they are either too small or too distant to be detected without amplification. Things can be hidden in time by being too fast for us to perceive or too slow for a single lifetime to encompass.

And, given our extraordinarily bandwidth-limited cognition and the fleeting nature of an individual life, it comes as no surprise that by far the majority of natural phenomena would be hidden.

It has been the great triumph of the sciences to find consistent means of studying natural phenomena hidden by both space and time—overcoming the limits of cognition and material culture. The scientific method is the portmanteau of instruments, formalisms, and experimental practices that succeed in discovering basic mechanisms despite the limitations of individual intelligence.

We are not interested in the expedients of spherical cows and point masses and frictionless surfaces or convenient idealizations that throw the complex baby out with the merely complicated bathwater.

There are, however, on this planet phenomena that are hidden in plain sight. These are the phenomena that we study as complex systems: the convoluted exhibitions of the adaptive world—from cells to societies.

Paradoxically the complex world is one that we can in many senses perceive and measure directly. Unlike distant stars or nearby minerals that require a significant increase in optical capability to arrive at insights into their elementary properties, behavior—both individual and collective—seems to present itself in a way that can be investigated rather modestly, through observation, or through experiment.

But the way in which complex phenomena are hidden, beyond masking by space and time, is through nonlinearity, randomness, collective dynamics, hierarchy, and emergence—a deck of attributes that have proved ill suited to our intuitive and augmented abilities to grasp and to comprehend.

The subjects of this volume are those worlds hidden by these attributes of complex systems. These are worlds that will be revealed not by better instruments but by new models and frameworks that allow us to see the familiar world in unfamiliar ways—to transform domains described into domains rigorously quantified and observations informally sensed into those formally understood.

Over the course of thirty years the Santa Fe Institute has been looking into this proximal, near-invisible reality. And looking not with multimillion-dollar facilities built for investigating the divisibility of the atom, but with pencil, paper, microcomputer, and collaborative minds. Working in highly diverse, nondisciplinary teams to invent new concepts to render up complex reality to science.

The last few decades represent a single-minded pursuit of tools and methods and frameworks that are adequate for exploring the adaptive world. We are not interested in the expedients of spherical cows and point masses and frictionless surfaces or convenient idealizations that throw the complex baby out with the merely complicated bathwater; the mission of the Santa Fe Institute is to search for order in the complexity of evolving worlds.

Reading over the nearly forty short essays making up this volume, it is interesting to observe a shift from an emphasis on what makes for invisibility (Mavericks), to the creation of new methods for seeing into complexity (Unifiers), culminating with the most recent instrumental use of theory to intervene into complex reality (Terraformers). From fundamental science to application, we never lose sight of the need to make these two necessary features of research work together.

1984–1999

MAVERICKS

COMPLEX ADAPTIVE SYSTEMS: A PRIMER

John H. Holland, University of Michigan
SFI Bulletin, Summer/Fall 1987

At the core of areas of study as diverse as cognitive psychology, artificial intelligence, economics, immunogenesis, genetics, and ecology, we encounter nonlinear systems that remain far from equilibrium throughout their history. In each case, the system can function (or continue to exist) only if it makes a continued adaptation to an environment that exhibits perpetual novelty. Traditional mathematics, with its reliance upon linearity, convergence, fixed points, and the like, seems to offer few tools for building a theory here. Yet, without theory, there is less chance of understanding these systems than there would be of understanding physical phenomena without the guidance of theoretical physics. What's to be done?

Hierarchical Organization and Building Blocks

There are some hints. First, all such systems exhibit a hierarchical organization. In living systems, proteins combine to form organelles, which combine to form cell types, and so on, through organs, organisms, species, and ultimately ecologies. Economies involve individuals, departments, divisions, companies, economic sectors, and so on, until one reaches national, regional, and world economies. A similar story can be told for each of the areas cited.

These structural similarities are more than superficial. A closer look shows that the hierarchies are constructed on a "building block" principle: subsystems at each level of the hierarchy are constructed by combinations of small numbers of subsystems from the next lower level. Because even a small number of building blocks can be combined in a great variety of ways, there is a great space of subsystems to be tried, but the search is biased by the building blocks selected. At each level, there is a continued search for subsystems that will serve as suitable building blocks at the next level.

A still closer look shows that, in all cases, the search for building blocks is carried out by competition in a population of candidates. Moreover, there is a strong relation between the level in the hierarchy and the amount of time it takes for competitions to be resolved—ecologies work on a much longer timescale than proteins, and world economies change much more slowly than the departments in a company. More carefully, if we associate random variables with subsystem ratings (say, fitnesses), then the sampling rate decreases as the level of the subsystem increases. As we will see, this has profound effects upon the way in which the system moves through the space of possibilities.

System–Environment Interaction

Common features of system–environment interaction in each case provide additional hints about the characteristics of the movement through the space of possibilities:

1. Each of the systems interacts with its environment in a game-like way: sequences of action ("moves") occasionally produce *payoff*, special inputs that provide the system with the wherewithal for continued existence and adaptation. Usually, payoff can be treated as a simple quantity (in physics, energy; in genetics, fitness; in economics, money; in game theory, winnings; in psychology, reward) sparsely distributed in the environment and that the adaptive system must compete for it with other systems in the environment.

2. The environment typically exhibits a range of regularities or *niches*, that can be exploited by different action sequences or *strategies*. As a result, the environment supports a variety of processes that interact in complex ways, much as in a multiperson

Chapter 1: Complex Adaptive Systems: A Primer

game. Typically, there is no superprocess that can outcompete all others, hence an *ecology* results (in physics, domains; in ecological genetics, interacting species; in economics, companies; in neurophysiological psychology, cell assemblies, etc.). The very complexity of these interactions assures that even large systems over long time spans can have explored only a minuscule range of possibilities. Even for much-studied board games such as chess and Go, this is true; the not-so-simply defined "games" of ecological genetics, economic competition, immunogenesis, central nervous system activity, etc., are orders of magnitude more complex. As a consequence, the systems are always far from any optimum or equilibrium situation.

> Even for much-studied board games such as chess and Go, this is true; the not-so-simply defined "games" of ecological genetics, economic competition, immunogenesis, central nervous system activity, etc., are orders of magnitude more complex.

---

3. There is a trade-off between *exploration* and *exploitation*. In order to explore a new niche, a system must use new and untried action sequences that take it into new parts (state sets) of the environment. This can occur only at the cost of departing from action sequences that have well-established payoff rates. The ratio of exploration to exploitation in relation to the opportunities (niches) offered by the environment has much to do with the life history of a system.

4. There is also a trade-off between "tracking" and "averaging." Some parts of the environment change so rapidly relative to a

PHASE I: MAVERICKS

given subsystem's response rate that the subsystem can react only to the average effect; in other situations, the subsystem can actually change fast enough to respond "move by move." Again, the relative proportion of these two possibilities in the niches that the subsystem inhabits has much to do with the subsystem's life history.

Pervasive Features of Subsystem Interactions

Beyond these commonalities, there are characteristic interactions between components that can be observed in each kind of system:

1. The value ("fitness") of a given combination of building blocks often cannot be predicted by a summing up of values assigned to the component blocks. This nonlinearity (commonly called *epistasis* in genetics) leads to coadapted sets of blocks (*alleles*) that serve to bias sampling and add additional layers to the hierarchy.

2. At all levels, the competitive interactions give rise to counterparts of the familiar interactions of population biology—*symbiosis, parasitism, competitive exclusion*, and the like.

There is an additional element of importance: these systems usually generate implicit internal models of their environments, models progressively revised and improved as the system accumulates experience. The systems *learn*.

---

3. Subsystems can often be usefully divided into *generalists* (averaging over a wide variety of situations, with a consequent

high sampling rate and high statistical confidence at the cost of a relatively high error rate in individual situations) and *specialists* (reacting to a restricted class of situations with a lowered error rate bought at the cost of a low sampling rate).

4. Subsystems often exhibit multifunctionality in the sense that a given combination of building blocks can usefully exploit quite distinct niches (environmental regularities), usually, however, with different efficiencies. Subsequent recombinations can produce specializations that emphasize one function, usually at the cost of the other. Extensive changes in behavior and efficiency, together with extensive adaptation, can result from recombinations involving these multifunctional founders.

Internal Models

There is an additional element of importance: these systems usually generate implicit internal models of their environments, models progressively revised and improved as the system accumulates experience. The systems *learn*. Consider the progressive improvements of the immune system when faced with antigens, and the fact that one can infer much about the system's environment and history by looking at the antigen population. This ability to infer something of a system's environment and history from its changing internal organization is the diagnostic feature of an implicit internal model.

The models encountered are usually *prescriptive*—they specify preferred responses to given environmental states—but, for more complex systems (the central nervous system, for example), they may also be more broadly *predictive*, specifying the results of alternative courses of action. We understand little of this process of model building, but it lies at the heart of the problems associated with the emergence of structure in complex systems. For processlike transformations, the relevant mathematical model is a *homomorphism*. Real systems almost never meet

PHASE I: MAVERICKS

the requirements for a homomorphism, but there are weakenings, the so-called *quasi-homomorphisms* (q-morphisms). The origin of a hierarchy can be looked upon as a sequence of progressively refined q-morphisms based upon observation.

Mathematical Concerns

In looking for a mathematics to deal with these commonalities, one finds relevant pieces in extant studies of particular examples. For instance, in mathematical economics, there are pieces of mathematics that deal with hierarchical organization, retained earnings (fitness) as a measure of past performances, competition based on retained earnings, distribution of earnings on the basis of local interactions of consumers and suppliers, taxation as a control on efficiency, and division of effort between production and research (i.e., exploitation vs. exploration). Many of these fragments, with due alteration of detail, can be used to study the counterparts of these processes in the other areas.

The most hopeful path seems to be a combination of computer modeling and a mathematics that puts much more emphasis upon *combinatorics* (the branch of mathematics dealing with combinations) and competition in parallel processes.

The task of theory is to explain the pervasiveness of these features by elucidating the mechanisms that assure their emergence and evolution. The most hopeful path seems to be a combination of computer modeling and a mathematics that puts much more emphasis upon *combinatorics* (the branch of mathematics dealing with combinations) and competition in parallel processes.

A prime objective of this theory should be an account of the emergence of q-morphisms in response to complex environments exhibiting sparse payoff. Computer simulations should give a better understanding of the conditions under which the phenomena of interest emerge. The close control of initial conditions, parameters, and environment made possible by simulation should enable the design of critical tests of the unfolding theory. And, as is usual in experiment, the simulations should suggest new directions for theory. The broadest hope is that the theoretician, by testing deductions and inductions against the simulations, can reincarnate the cycle of theory and experiment so fruitful in physics.

6

BOUNDED RATIONALITY AND OTHER DEPARTURES

*George Cowan, SFI, interviews
Kenneth Arrow, Stanford University
SFI Bulletin, Winter/Spring 1989*

GEORGE COWAN: We are eager to know what fresh ideas there are in economics. What particularly intrigues you? What directions are things moving in? What stimulates change? What does the future hold?

KENNETH ARROW: The next big steps forward involve consideration of nonlinear dynamics and future uncertainties. This is the warp and woof of a lot of economic research.

First, let me review some current ideas. The standard picture of the economy is one which assigns a very great weight to the system of prices for directing the way in which the economy functions.

Firms and households are assumed to be rational in the sense that they are trying to maximize something—in the case of firms, their profits. A firm has various ways of combining inputs to produce possible combinations of outputs. Each technologically feasible combination will produce a certain profit at given prices. The firm, therefore, will choose that set of inputs and outputs that maximizes profit. As prices change they choose different combinations and different scales.

Similarly, the household, which is the consumer, is thought of as choosing some bundle of inputs and outputs, the outputs usually being labor, but possibly other commodities that they own, in order to achieve as high a level of satisfaction or utility as is possible. This procedure defines for each firm or household a certain supply and demand for each commodity. The demand for any commodity or

PHASE I: MAVERICKS

the supply will depend not only on the price of that commodity but on the prices of all other commodities. If nothing else, the rise in the price of some other good will reduce the amount of purchasing power available to the consumer for the purchase of this good. In addition, the price of gasoline, for example, affects the extent to which people use automobiles. The result is that for every set of prices there are a lot of supplies and demands that can be added up over all agents in the economy, giving rise to a net demand. An equilibrium set of prices means those prices at which net demand goes to zero for all commodities and supplies and remain balanced.

So prices are a result of demands and supplies reacting to create an equilibrium in each situation. Here we are assuming a process by which a set of prices that are not in equilibrium can be altered. We are assembling a great deal of information, which is dispersed throughout the economy in this process. Defects in this information can trigger the statement that the economy is out of equilibrium. It is somewhat more problematic to determine how, with good information, we can guide the economy from an out-of-equilibrium position to equilibrium. This is still an open question.

The present analysis goes further. Obviously, a great many economic decisions are made with a view to the future. For example, when one buys a house, one is expecting to use that house in the future; the ability to use that house will depend upon the ability to pay for the electricity, gas, and so forth. Similarly, in the case of a business firm, expenditures are made, that is, investments in plant and equipment are made. These are inputs to produce future output and are justified only to the extent that output in the future will eventually cover the relevant costs. In an idealized model of an economy, we now have markets for all future goods. Thus, an automobile manufacturer who constructs and equips a plant is, at the same time, selling the product of that plant—automobiles produced five years hence. Therefore, in this ideal model the same structure given to the current-equilibrium case applies.

I shall mention two implications. For one, many markets have to clear. In each the supply has to equal demand. Second, this process gives rise to a path in quantities and to a path in prices. It has usually been assumed, without analysis, that these paths tend to converge to a stationary state. Somewhat more exactly, in light of the fact that both technology and population are increasing, they converge to a steadily growing path.

To complete this picture of the ideal model, we have to allow for uncertainty. For example, innovations will take place in agriculture and we don't know what the weather will be. So, the ideal model goes one step further and says there should be a separate market for every future good for each possible history of contingencies.

Again, we are multiplying the number of markets. This picture is very complex. We immediately notice that many markets are being assumed that do not, in fact, exist. We do not have many ways of contracting to deliver goods in the future, except in very limited markets, such as agricultural commodities. We certainly have no great number of markets for results contingent on events, although they exist: insurance policies are the most obvious example.

But most of the markets called for by this idealization are nonexistent. Sometimes they can be replaced by a chain of other markets. To some extent this happens, but very clearly the economy, in its intertemporal and risk-bearing aspects, is not fully regulated by markets that exist, even in combination. Therefore, a notion that has played an effective role in both theory and practice is that economic agents anticipate correctly what prices would be in markets that don't exist. Notice this doesn't mean, for example, that the price of wheat five years from now is supposed to be anticipated correctly; of course, we know that there are contingencies. But we

PHASE I: MAVERICKS

are saying that, for any given set of contingencies, we can predict what the price of wheat will be.

In the end, we still have a lot of problems. We have a problem of how prices even come to work. What is the process by which they make the market? The suggestions I've made turn out mathematically not to imply convergence, in general, except under a very restrictive hypothesis that we do not want to make. We have a problem that the standard model requires incredibly many markets, and the substitute model requires incredible recalculation. The suggestion, therefore, is that we have to emphasize a different kind of world, one in which people, instead of optimizing and rationally forecasting the future, are engaged in much more limited operations more suitable to constraints on human reasoning and calculating abilities and even to those abilities as augmented by computers.

As a new point of view we turn to bounded rationality, a departure from the mainstream tradition. We no longer can assume that every agent is a perfect calculator. This point of view is given a great deal of emphasis by Herbert Simon. Simon argued that people do not maximize. When they're forecasting the future, they do not perform the task of rational analysis. Modern cognitive psychology is almost dominated by emphasis on irrationality in the formation of beliefs and expectations, and experiment after experiment has confirmed this.

G. COWAN: In this view, does a very large number of participants in a given market expand the space in which the calculation occurs? In other words, is the operation of a million brains additive in some sense and mutually supportive?

K. ARROW: My answer to that is unequivocally yes and no. Yes, because it is true that different people can explore different parts of the universe and the competitive struggle would suggest that those who do the best exploration will triumph in the end. If the result of the exploration is to do something more cheaply, then that will drive the others out of business or cause them to imitate it so the idea will spread.

On the other hand, many people spend most of their creative power in trying to figure out what other people are doing.

G. COWAN: And I suppose they're also influenced by what the other guy is doing, the herd instinct.

K. ARROW: One possibility is that people don't innovate enough because, if most people aren't innovating, they assume noninnovation must be the right thing. Then you have a second-order effect that says, if everybody is doing something, I'm not going to make any money doing it that way. The only hope is to do something different. So it might work in either direction. We are trying now to use tools which are developed in other disciplines but which deal with somewhat analogous problems. We do not expect these tools are transplantable without change because they were developed to address a different set of phenomena. The intertemporal equilibrium path, as I have suggested, is supposed to converge toward a steady state or steady growth. However, we can make all the assumptions with perfect foresight, well-behaved production-possibility sets, and so forth, and still it turns out that other behavior, including chaotic behavior, is possible.

G. COWAN: Even cyclic behavior, I suppose...

K. ARROW: Cyclic behavior or chaotic behavior, yes. All these have been exhibited within models that are highly classical. This is more undermining than might appear. It isn't merely that we get a bigger repertory of solutions. The chaotic solutions are disturbing because it's incredible that you can have perfect foresight along a chaotic path. The existence of a chaotic solution, in a sense, undermines the idea of perfect foresight. This points to an interesting way in which economic analysis differs from physics. In economics the particles, that is, agents, are endowed with some kind of foresight. Their

PHASE I: MAVERICKS

image of the future affects the present. But experienced people are not entirely crazy, and their image of the future and the real future may not be entirely different. People do have some idea of what's going to happen. Their idea of what's going to happen affects what they do today; what they do today affects what's going to happen. So one really unsolved problem has to do with including limited foresight. We have some idea how to model perfect foresight. We have very little way of knowing how to model limited foresight, even to define the term. I think that the dynamic analysis will have to be modified to account for foresight. It is already in our present models but it is treated as perfect, which is not realistic.

G. COWAN: I assume that foresight, however limited, will tend to be self-fulfilling.

> We have ice ages and hot periods in economics, new innovations, foreign competitors, and all the rest of the things in the world that change and evolve.

K. ARROW: Not necessarily, but that is frequently the case. The opposite is also sometimes true. Robert K. Merton produced the idea of a self-fulfilling prophecy and a self-denying prophecy. A classic example has to do with elections. Suppose we have polls showing that "A" is going to win. One possible consequence is that the supporters of "A" don't bother going to the polls, while supporters of "B" may double their efforts. Of course, the opposite is also possible. Other examples: Suppose a stock's price has been rising steadily. If tomorrow's price is the same as today's, I might assume the stock price has stopped rising. Then I'll get out of the

Chapter 2: Bounded Rationality and Other Departures

stock and that, of course, will cause the price to fall. An important reason for holding stocks is that they will continue to provide a good rate of return in capital gains. Any assumption that will make them reach a peak may not cause the peak to be maintained but may cause it to go down.

People operate in a complex environment. Rationality enters at two levels here, first at the level of forming a model and using it for prediction, and second at the level of behavior. At the first level you just don't have enough experience to really get a whole model; you only have a limited set of observations. You can't even use those efficiently due to limitations beyond your control. So, we find that biologists and computer scientists inspired by biologists have been discussing the idea of getting better and better, not all at once, but step by step. We can discuss several different levels. We can consider a landscape in which we are trying to climb a fitness peak. The problem is that there may be more than one peak. In fact, even if there's only one peak, it would be pretty hard to climb if it is complicated. On top of that, we have the fact that I am gradually adjusting to conditions around me, and so are all the other economic agents in the systems; this may correspond to the idea of coevolution. Our objective is moving around because other people are involved. And finally, in the economic world and, indeed, in the biological world, the outside environment is changing. We have ice ages and hot periods in economics, new innovations, foreign competitors, and all the rest of the things in the world that change and evolve.

So, we must include the question of learning, of adaptation, and the fact that the agents influence each other, and that the underlying environment can change. We'd like to think that such concepts come back to the very beginning [of this discussion] when we spoke of adjusting prices until they come to equilibrium, so that the process becomes a learning process. We are reluctant to

PHASE I: MAVERICKS

entirely abandon the idea of *intertemporal equilibrium*. So we are trying to borrow and modify these new tools to incorporate a kind of dynamic analysis that goes beyond what is already in economics. The variety of behavior that these processes can exhibit has not been understood until the last few years. Another new factor is further development of the role of learning feedback. How do people behave, what simple response mechanisms are available to them in order to adjust to changing information?

G. COWAN: In intertemporal equilibrium, there are various factors influencing the particular level of equilibrium. Some kinds of information are available very rapidly and other kinds more slowly. The various factors may change at very different rates. What are the appropriate timescales?

K. ARROW: If the information disseminates rapidly, prices may change rapidly.

G. COWAN: In that case, what would we mean by intertemporal equilibrium?

K. ARROW: In the new scheme there will not be intertemporal equilibrium, but there will be some concept that takes the place of that which changes all the time.

G. COWAN: Let's go back and review the fresh breezes in economics. You introduced bounded rationality and ...

K. ARROW: Also genetic algorithms, biological analogies, and fitness landscapes.

G. COWAN: What we're talking about in all of these cases is a learning and adaptive process.

K. ARROW: Well, bounded rationality is kind of a constraint on the learning process. I also want to draw from cognitive psychology

Chapter 2: Bounded Rationality and Other Departures

because we've developed certain patterns that tell us how people actually behave.

G. COWAN: But, in general, what you're saying is that there is a process going on among these agents by which everybody learns, remembers, and adds to the total stock of information which is being applied to a particular situation.

K. ARROW: Well, the memory aspect is interesting. The genetic algorithm, if you take it literally, is static. At any moment you've just got a given set of rules. One thing the genetic algorithm doesn't do is remember the path it takes. New observations cause you to modify rules. Those observations are no longer available. They're incorporated in the packaging. If we have settled on a new set of rules, in some sense the past is not relevant. If some brand-new rule came into the set, one thing we might do is go back over the past and consider how well it might have worked in the past. It wouldn't be difficult to add except that it would impose tremendous memory demands.

G. COWAN: But the human memory is capable of that kind of thing. It may operate in precisely that way.

K. ARROW: There are some approaches that depend upon using the whole past. There are other approaches in which you start where you are now and never mind how you got there. In the price adjustment, the simplest rule is, if demand exceeds supply, raise the price in that market. You're looking only at the present situation. It's true the prices you're starting with are the result of the whole past, but you use only the present. You forget about the rest of your story. Even though, theoretically, you could go back and look at the points on that demand function and ask if it is a plausible demand. So there are, so to speak, two schools of thought, one of which says you start with a certain approximation and then you go on. Another group says you never leave the past out.

PHASE I: MAVERICKS

In a two-person game, there's an outcome. We both know the payoffs, so I observe his strategy, and I'll pick a strategy which is optimal against his. Having played many times, I'll note that he's made strategy choices. I now assume he's going to use these strategies with probabilities proportional to their frequencies. So at any moment all the past elements appear in the summary form. But now you're throwing out some information; you're not using the whole past because all you're using is the summary.

G. COWAN: You could save something. You could date every input.

K. ARROW: You could do that, but is there a way you could effectively use those data?

G. COWAN: I think the mind does date and otherwise identify data inputs and weights each input by some set of rules that uses this kind of information.

K. ARROW: Perhaps so. I would tend to say that the near past is more likely to be relevant than the distant past.

G. COWAN: I think we do that. Conversely, we might have reason to discount the near past.

K. ARROW: You're moving down the slippery path toward rationality. When you say the mind is that intelligent, you're implying really rational analysis.

G. COWAN: But the mind tends to operate that way.

K. ARROW: We know it doesn't do it well. The psychologists find there are some circumstances when you are very conservative, and other times when you overreact to current data. People can ignore basic data and past experience and react too much to the current situation. Sometimes they presume a long history summed up in one fact, but there's a tendency to look only at the latest information and never mind the long term. People also tend to be quite indifferent to the size of the sample. They feel just as strongly about it if it's two to one as if it's twenty to ten. Yet, obviously, any theory you can think of would suggest that you ought to be a lot more sure about a fact when it's twenty to ten than when it's two to one. So there are certain biases; some tend to overemphasize the present relative to the past, and some tend to develop inflexibility.

G. COWAN: We touched on many interesting questions but left some out, ethical standards, for instance. And we barely touched on some of the properties that are apparently common to all complex systems, like multiple solutions rather than unique solutions, and rarely, if ever, best solutions. Perhaps we should leave these questions for another time, not too far in the future, I hope. Thank you very much.

CAN PHYSICS CONTRIBUTE TO ECONOMICS?

Richard Palmer, Duke University
SFI Bulletin, Winter/Spring 1989

Physicists are not known for their humility. There is often the feeling among them that physics is the hardest subject around, at least among the quantifiable sciences, and that a sound training in physics is a license for tackling well-posed problems in *any* area. Why, then, is economics still at large, despite the present interactions between the fields? Are the problems actually *hard*, or are they not well posed, or is the arrogance unjustified? The answer should surely be "all three," although it should also be said that physicists do not have a monopoly on arrogance.

The real issue is the degree of similarity between the fields. The underlying supposition of the dialogue between natural scientists and economists arranged by the Santa Fe Institute is that there is enough common ground to make progress possible. Economic systems can be seen as complex, nonlinear, interacting systems of many parts, and this description also fits problems actively studied in the natural sciences. But, on a more detailed level, is there really enough similarity with economics to justify optimism? It is crucial to go beyond generalities and ask just how the fields differ, and how they are alike.

Dynamical Systems vs. Foresight

Most situations in physics involve particles or objects (e.g., wave functions in quantum mechanics) whose future behavior may be predicted from a knowledge of the present. No knowledge of—or expectation about—the future is needed. In contrast, the regnant approach in economics, rational-expectations theory with foresight,

PHASE I: MAVERICKS

relies explicitly on economic agents who can anticipate the future; the "particles" of the theory form strategies on the basis of future expectations. At first sight, this difference is central, but on closer examination it may not be so serious.

On the one hand, there *are* situations in physics where the particles effectively explore possible futures to determine their behavior. Quantum mechanics and classical mechanics can both be expressed in terms of minimizing a "path-integral" quantity across all possible future histories. In each case, however, the theory can also be formulated in a way that avoids the apparent dependence on the future. Could the same be done in economics? Can the path-integral methods developed in physics be applied to economics?

The laws of physics do not seem to improve with the universe's experience.

On the other hand, the foresight approach is in trouble in economics. The simplest way of formulating the theory expects the agents to be infinitely smart and well informed, a condition clearly not fulfilled in the real world. Economists are now trying to construct modified "bounded rationality" theories that require only limited intelligence or computational power in the agents. Though promising, these ideas are seen as less attractive because, while there is often only one way to be perfectly rational, there are many forms of irrationality, and uniqueness becomes a major issue. An approach in which many of the SFI participants are involved is to give agents only very limited ability to anticipate the future but to let them *learn* as time progresses so as to optimize their predictions.

Chapter 3: Can Physics Contribute to Economics?

A more ambitious type of learning, also under study by SFI participants, is the adaptive systems approach. Here, the agents again learn with experience (though not necessarily explicitly anticipating the future), in a rule-based framework. Wholly new rules or modes of behavior can evolve, and the system can adapt to a changing environment. New strategies may in fact coevolve while interacting with each other. These ideas are linked more closely to biology than to physics; the laws of physics do not seem to improve with the universe's experience.

Linearity vs. Nonlinearity

Linearity should not be an issue. Economic systems are obviously nonlinear, as are many, if not most, systems of current interest in physics. A more controversial question concerns the direction of feedback. Whereas a strictly linear system can have only negative feedback if divergence is to be avoided, positive feedback can occur in nonlinear systems if a saturation mechanism operates. Such systems tend to have multiple equilibria or resting points and great sensitivity to initial conditions. Traditionalists find it hard to relinquish uniqueness and global stability, but physicists are easily convinced and find positive feedback natural.

Deterministic vs. Stochastic Systems

Realistic economic theories have to admit the existence of external noise or "exogenous shocks" coming from factors not modeled economically. The dynamics is thus stochastic, involving chance. Physicists are too used to including random noise in their equations, usually to represent the effect of thermal fluctuations. In physics, a crucial question is the autocorrelation of the noise; uncorrelated or "white" noise is easily dealt with, but correlated noise can be very tricky. The noise sources in economics include psychological, social, and political effects in the population (unless these are modeled

PHASE I: MAVERICKS

explicitly, which is rare) and are surely not uncorrelated. Are the white noise theories therefore invalid?

Even completely deterministic systems can appear to behave rather randomly when the solutions are "chaotic." It has been popular recently to look for such deterministic chaos in economics, but the fad now seems to be dying down, as perhaps it should. It is certainly true that nonlinear economic models can have chaotic solutions, but it is doubtful whether these occur if the models are adjusted to match reality. And the analysis of economic time series to search for chaotic attractors is plagued by woefully insufficient data and can rarely be taken seriously. Chaos *may* occur microscopically, but it is unlikely to occur collectively or to be visible macroscopically in aggregated time series. Could one see it by examining many time series in parallel?

> **It is generally assumed by economic theorists that a model must be fully specified at the microscopic level; the strategies of particular agents must be known in detail. There was almost shock when an "agent-free" theory was suggested at SFI.**

Spatial Structure

Almost all theories in physics involve the spatial as well as the time domain. Many phenomena with possible parallels in economics require an interpretation for different spatial locations. Certainly, one can use geographical location, but it may not make the most productive analogy. Different *economic sectors* seem likely to be a better choice. In physics, a crucial determinant of behavior is very often the spatial dimensionality, with values other than three often being considered for theoretical reasons. The effective dimensionality for a network (or "web") of interconnected economic sectors is less easy to determine. It depends in general on the range of the interactions between sectors: Does everything interact strongly with everything else, or only with a smaller number of related sectors? If the effective dimension is high enough, then "mean field" methods from statistical mechanics may work well in economics.

Microscopic Details vs. Generic Behavior

It is generally assumed by economic theorists that a model must be fully specified at the microscopic level; the strategies of particular agents must be known in detail. There was almost shock when an "agent-free" theory was suggested at SFI. And yet some very successful theories in physics have thrown away most of the microscopic details in a problem and kept only a few central features, average properties, or statistical distributions. One important lesson of the last few decades of work in statistical mechanics has been the realization that there are many behavioral properties for which microscopic details do *not* matter. There may be just a few "universality classes" of behavior, in each of which the simplest possible examples is adequate for describing the generic behavior. Might not the same be true in economics?

Can we imagine theories that take into account little more than the topological or statistical structure of the web of interactions between economic agents or sectors? A number of SFI participants think this is possible and are investigating several directions. One intriguing idea is the analog of "self-organized criticality" in physics: certain stochastic dynamical systems evolve naturally toward a "critical" state in which the distribution of both length and timescales has a special "power law" form. Models in which spatial diffusion only occurs when a threshold is exceeded fall into this class, and it's

PHASE I: MAVERICKS

easy to find these conditions in economics. A transition to a higher technology does not diffuse from firm to firm, or from sector to sector, until the pressure for change exceeds a threshold dependent on setup costs and other factors. If the self-organized criticality paradigm applied, one would expect to see scaling behavior in economic time series and in the distribution of inhomogeneity. A detailed analysis will require knowledge of the statistical structure of economic webs, perhaps available partially in the structure of input/output matrices. Of course, the whole web structure may itself evolve in time.

Simple vs. Complex Systems

Complex system probably has as many definitions as there are practitioners, but one common theme is the existence of many possible stable or metastable states (or equilibria), as found in glasses, spin glasses, neural networks, and such. A "rugged landscape" picture is often invoked for the energy as a function of the many-dimensional state of the system. The multiple states and ruggedness usually come about through "frustration," the impossibility of satisfying many conflicting constraints simultaneously. Are there parallels in economics, and are economic systems complex in this sense?

Certainly, many economic theories have multiple solutions, though this is often regarded more as a defect than as an expected feature. The natural scientists find it more natural. One might also expect to find a degree of frustration in economic systems, where, for example, trading in an optimal fashion between pairs of agents might not satisfy more global constraints. The reverse might also occur, in which larger markets allow better trading for all. In any case, neither frustration nor rugged landscapes have been much investigated in economics.

Conclusion

I have described a number of similarities, and some major differences, between problems in physics and in economics. I have also posed many unanswered questions: there is no lack of open problems! Since no major breakthroughs have yet emerged from the meeting of the fields, it is clearly too soon to call the endeavor a success. But there seem to be plenty of reasons for optimism, plenty of interesting ideas needing further exploration, and plenty of exciting challenges for all involved.

Addendum

SIMPLE PROBLEMS, COMPLEX SOLUTIONS

Suppose that there are strong social norms as to what type of vehicle to drive in Santa Fe. There are, in actuality, only two available types of vehicles in town: cars and pickups trucks. The social pressure is such that it overrides any cost differences between them and, if over one half of the population own a certain type of vehicle, you prefer to own that type. Every now and then, vehicles break down and must be replaced. What will be the characteristic Santa Fe vehicle given each of the following initial circumstances: everyone last year drove cars; everyone drove pickup trucks; or half the people drove each type of vehicle?

Even such a simple economic problem may have difficult solutions. There are two likely outcomes: a heavy predominance of cars or a heavy predominance of pickup trucks. The problem is how does the system arrive at one solution rather than the other? What if both possibilities are equally likely? What if pickup trucks are much cheaper than cars, yet everyone last year drove cars? Is it possible for everyone to be "locked" into a technology that is inferior, too expensive, or otherwise undesirable?

To extend this example to other economic events, one might consider Betamax versus VHS standards in the VCR market: has

PHASE I: MAVERICKS

the inferior product captured the predominance of the market? In modeling economic development, the two options could represent primitive and modern factories.

The potential dynamics of the above scenarios provide a clear contrast between the physicists' and economists' solutions. In a physical system, it would be adequate to take the goals of the agents (which option to choose) and the current state of the system (which option is predominant), and allow the agents to act myopically, that is, to follow their goals assuming that everyone else will continue doing what they are doing. However, an economist would suggest that economic agents may be endowed with expectations and strategic motivations. Agents may base their choices not only on what other agents are currently doing, but also on what the agents think the others will be doing, and how their actions will affect the actions of others.

These scenarios also illustrate the notion of linkages in economics. Some goods may depend on other goods for their existence (for example, internal combustion engines depend on the availability of gasoline). Moreover, the development of new goods may create and destroy different industries (for example, cars created the market for service stations but destroyed most of the buggy whip industry). So, how does one choose?

LEARNING AND GAMES

The analysis of games is important to a broad spectrum of disciplines, including economics, biology, political science, and sociology. This problem is an example of game theory.

You have arranged to meet a friend for lunch, but you forgot to specify at which restaurant. Unfortunately, there is not enough time to contact one another. In the past you have always met at two popular eateries, Restaurant X and Restaurant Y. You both would rather eat together than apart. To which restaurant do you go?

Chapter 3: Can Physics Contribute to Economics?

This is a game of coordination. The traditional approach to solving it is to see whether there is a way to prescribe choices such that if both players knew the prescription, they both would follow it. For example, suppose that both you and your friend receive a message that says to meet at Restaurant X. Given that your friend is going to be at Restaurant X, then it is in your best interest to go there. Similarly, if you are going to Restaurant X, then it is in your friend's best interest to meet you there. Thus both of you meeting at Restaurant X as a result of the message is a solution to this game. (This remains a solution even if both of you would rather eat at the other restaurant.)

The main problem with the prescriptive approach is that it makes sense only if both parties know the prescription. An alternative approach is to incorporate learning models into each agent's behavior. What if you and your friend often have similar lunch plans? The experience gained over past meetings could have important implications for where you will go this time. What if one restaurant has an outdoor patio, and it is warm outside?

Another important element of games is that one's best move often depends on the strategies that the opponents choose. Thus a very interesting coevolving system emerges. The best move depends on the opponent's moves, which in turn depend on one's own move. Simply performing better than one's opponents may be as important as performing the best one possibly can, in these systems.

—*John Miller, SFI Bulletin Winter/Spring 1989*

NATURE CONFORMABLE TO HERSELF

Murray Gell-Mann, Caltech
SFI Bulletin Winter/Spring 1992
From a talk given at the 1992 Complex Systems Winter School,
Tucson, Arizona

More than thirty years ago, I was the first visiting professor at the College de France in Paris, with an office in the laboratory of experimental physics. I noticed that my experimental colleagues were frequently drawing little pictures in their notebooks, which I assumed were diagrams of experimental apparatus. But it turned out that those drawings were mostly of a gallows for hanging the vice director of the lab, whose rigid ideas drove them crazy.

I got to know the *sous-directeur* and talked with him on various subjects, one of which was Project Ozma, the attempt to detect signals from another technical civilization on a planet of a nearby star. SETI, the Search for Extraterrestrial Intelligence, is the present-day successor of that project. "How could you communicate if you found such a civilization?" he asked, assuming both interlocutors would have the patience to wait for the signals to be transmitted back and forth. I suggested that we might try beep, beep-beep, beep-beep-beep, for 1, 2, 3, and so forth, and then perhaps 1, 2, 3 . . ., 92 for the stable (except 41 and 63) chemical elements, etc., etc. "Wait," said the *sous-directeur*, "that is absurd. The number 92 would mean nothing to them . . . why, if they have 92 chemical elements, then they must also have the Eiffel Tower and Brigitte Bardot."

That is how I became acquainted with the fact that French schools taught a kind of neo-Kantian philosophy, in which the laws of nature are nothing but Kantian "categories" used by the *human mind* to grasp reality. (Many also taught, by the way, that artistic criticism is absolute and not a matter of taste, while the opinion that artistic standards are relative was treated as a feature of Anglo-Saxon pragmatism.)

PHASE I: MAVERICKS

Another notion of a quite different kind, far more platonic, is rife in mathematical circles in France (and elsewhere). That is the idea that the structures and objects of mathematics—say, Lie groups—have a reality, that they exist, in a sense, somewhere beyond space and time. (It is easy to see how one can come to think that way. Start with the positive integers—they certainly exist, in the sense of being used to count things. Number theory—okay. Zero and negative numbers—why not? Fractions, square roots? Solutions of algebraic equations in complex numbers? Probably— one is on a slippery slope.) These two points of view are argued in a book, *Matière à Pensée*, published recently by the biologist Jean-Pierre Changeux and the mathematician Alain Connes. I shall not inflict all their philosophical arguments on this congenial group, and anyway I have never studied them carefully. Let me say merely that the authors do raise the question of what is the role of mathematical theory in our understanding of the world, especially the physical world.

I like to put the relevant questions in the following form: Would advanced complex adaptive systems on another planet come up with anything like our mathematics or anything like our mathematical theories of physical processes, or both? At present, we can only speculate about the answers, but the questions are deep and meaningful.

Eugene Wigner once wrote an article titled "The Unreasonable Effectiveness of Mathematics in the Natural Sciences." I don't know what he wrote in the article, but it is certainly a fact that, up to now, especially in the domain of fundamental physics, we have had striking success with our use of mathematics.

Sometimes, as with Fourier series, the physicist has to invent the mathematical trick and the mathematicians later formalize and adapt it. Sometimes, as with Heisenberg and matrices, the concept is already known to mathematicians and physicists but not to the particular theoretician involved; he reinvents it. Often, as with

Chapter 4: Nature Conformable to Herself

Einstein, the physicist senses what he wants and asks a mathematician to provide it—in the case of the equation describing general relativistic gravitation, Einstein asked his old classmate, Marcel Grossmann, for the tensor he needed, and thus the Ricci tensor became the Ricci–Einstein tensor.

More recently, abstract mathematics reached out in so many directions and became so seemingly abstruse that it appeared to have left physics far behind, so that, among all the new structures being explored by mathematicians, the fraction that would even be of any interest to science would be so small as not to make it worth the time of a scientist to study them.

But all that has changed in the last decade or two. It has turned out that the apparent divergence of pure mathematics from science was partly an illusion produced by the obscurantist, ultrarigorous language used by mathematicians, especially those of a Bourbakian persuasion, and by their reluctance to write up nontrivial examples in explicit detail. When demystified, large chunks of modern mathematics turn out to be connected with physics and other sciences, and these chunks are mostly in or near the most prestigious parts of mathematics, such as differential topology, where geometry, analysis, and algebra come together. Pure mathematics and science are finally being reunited and, mercifully, the Bourbaki plague is dying out. (In the late Soviet Union, they never succumbed to it in the first place.)

An anecdote will illustrate the situation during the '50s. In 1955, at the Institute for Advanced Study in Princeton, Frank Yang was discussing with other physicists the recently developed Yang–Mills quantum field theory. At the same time, S. S. Chern was lecturing on pure mathematics. Not only did Frank attend some of the lectures, but he and Chern were old friends, their children played together, and Chern had been one of Frank's teachers in China; neither of them noticed that Chern's lectures on fiber bundles were basically

PHASE I: MAVERICKS

concerned with the same subject as Frank's lectures on Yang–Mills theory! In fact, they didn't learn about this equivalence for many years.

Yang–Mills theory, from the physics point of view, was a generalization of quantum electrodynamics from the gauge group U_1 to the gauge group SU_2 with noncommuting charges. Later, we generalized it to all products of U_1 factors and simple compact Lie groups, including SU_3. Today the "standard model" of elementary particle physics, apart from gravitation, is based on $SU_3 \times SU_2 \times U_1$. Moreover, Einsteinian gravitation has strong parallels with generalized Yang–Mills theory, although the gravitation theory is based on the noncompact Lorentz group and involves a tensor instead of a vector field. What does it mean that this progression from one gauge group to another has worked so well? Are we really dealing with something peculiar to the human mind or with a phenomenon so deeply rooted in the properties of nature that any advanced complex adaptive system would be likely to follow similar paths?

A related set of issues was discussed more than three hundred years ago, especially by Isaac Newton. Children learn that he thought of the theory of universal gravitation when an apple fell on his head. Well, not on his head, but nearby, anyway.

Historians of science are not sure whether to credit the apple at all, but they admit that there could have been an apple. As you know, in 1665 the University of Cambridge closed up on account of the plague and sent everyone home, including Newton, a fresh BA, who went back to Woolsthorpe, Lincolnshire. There, during 1665 and 1666, he thought a little about integration and differentiation, a little more about the law of gravitation, and a lot about the laws of motion. Moreover, he carried out the experiment showing that white light is made up of the colors of the rainbow. While historians of science now emphasize that he didn't completely clear up all these matters in one *annus mirabilis*, or "marvelous year," they admit that he made a good start on all of them

Chapter 4: Nature Conformable to Herself

around this time. As my friend Marcia Southwick says, he could have written a pretty impressive essay on "What I Did During My Summer Vacation."

As to the apple, there are four independent sources. One of them, Conduitt, writes:

> In the year 1666 he retired again from Cambridge . . . to his mother in Lincolnshire & whilst he was musing in a garden it came into his thought that the power of gravity (wch brought an apple from the tree to the ground) was not limited to a certain distance from the earth but that this power must extend much farther than was usually thought. Why not as high as the moon said he to himself & if so that must influence her motion & perhaps retain her in orbit, whereupon he fell a calculating what would be the effect of that supposition but being absent from books & taking the common estimate in use among Geographers & our seamen before Norwood had measured the Earth, that sixty English miles were contained in one degree of latitude on the surface of the Earth his computation did not agree with his theory & inclined him then to entertain a notion that together with the force of gravity there might be a mixture of that force wch the moon would have if it was carried along in a vortex . . . .

What interests us here is the extrapolation—if gravitation applies on earth, why not extend it to the heavens and use it to explain the force that keeps the moon in its orbit? Here is how Newton describes the idea much later:

> How the great bodies of the earth Sun moon & Planets gravitate towards one another what are the laws & quantities of their gravitating forces at all distances from them & how all the motions of those bodies are regulated by

PHASE I: MAVERICKS

those their gravities I shewed in my Mathematical Principles of Philosophy to the satisfaction of my readers: And if Nature be most simple & fully consonant to her self she observes the same method in regulating the motions of smaller bodies which she doth in regulating those of the greater. This principle of nature being very remote from the conceptions of Philosophers I forbore to describe it in that Book least I should be accounted an extravagant freak & so prejudice my Readers against all those things which were the main designe of the Book.

Today some of us have the same concerns about extrapolation as the ones to which Newton refers—Shelly Glashow inveighs against superstring theory because, in embracing Einsteinian gravitation, along with the other forces, it achieves its synthesis around the Planck mass of 2×10^{19} GeV, larger by a gigantic factor than any energy at which particle physics experiments are carried out. But he and others were in the forefront of extrapolating the standard model $SU_3 \times SU_2 \times U_1$ to a unified Yang–Mills theory based on SU_5, in which the unification (without gravitation) is achieved around 10^{14} or 10^{15} GeV, which lies most of the way to the Planck mass. Moreover, Shelly and others have alleged that nothing much could happen in between present energies and 10^{14} GeV or so—there would just be a desert.

Well, here in Arizona we know that deserts are not necessarily empty, and that they are often very rich in plant and animal life, so the gap between our experimental energies of a few hundred GeV and 10^{14} GeV may well contain some interesting flora and fauna, especially the supersymmetric partners of the known particles—as suggested by superstring theory. In fact, the unified super-Yang–Mills extrapolation works much better than the straight unified Yang–Mills theory (what some people call, quite inappropriately, in my opinion, "grand unified theory").

Chapter 4: Nature Conformable to Herself

But, to return to Newton, he was not thinking only of extrapolation. He returns repeatedly in his writings to the idea that Nature is consonant and conformable to herself in more general ways. From the *Opticks*:

> For Nature is very consonant and conformable to her self . . . . For we must learn from the Phaenomena of Nature what Bodies attract one another, and what are the Laws and Properties of the Attraction, before we enquire the Cause by which the Attraction is perform'd. The Attractions of Gravity, Magnetism, and Electricity, reach to very sensible distances, and so have been observed by vulgar Eyes, and there may be others which reach to so small distances as hitherto escape Observation; and perhaps electrical Attraction may reach to such small distances, even without being excited by Friction.

Thus, he thought of the laws as exhibiting conformability among themselves as well as within each one, just the kind of idea that we have followed in going from electrodynamics to QCD and the electroweak theory and then onward to unified Yang–Mills theory and, with gravitation included, to the superstring theory.

If we modernize Newton's conception a bit, we could say that the laws of nature exhibit a certain amount of self-similarity and not, of course, perfect scaling, but rather the kind of thing one sees in the Mandelbrot fractal set. So, in peeling the skins off the onion of fundamental physics, we encounter certain similarities between one layer and the next. As a result, the mathematics with which we become familiar on account of its usefulness in describing one layer suggests new mathematics, some of which may apply at the next layer—in fact, even the old mathematics may still be useful at the next layer. These generalizations may be performed either by theoretical physicists or by mathematicians. If pure mathematicians are exploring ambitious generalizations of known mathematical

PHASE I: MAVERICKS

structures, they will surely run across some of the new ones that are needed—along with much more besides.

Ultimately, then, we can argue that it is the self-similarity of the structure of fundamental physical law that dictates the continuing usefulness of mathematics. Suppose that the fundamental theory of the elementary particles and their interactions is really heterotic superstring theory. It has a huge set of symmetries, including the conformal string symmetries that encompass the bootstrap principle and general relativity, and an internal symmetry group E8 x E8 that undergoes spontaneous symmetry breaking. The outer layers of the onion show gravitation and electromagnetism. Penetrating a little further turns up SU2 x U1, SU3 of color, and the bootstrap idea. And so it goes on. The mathematics at each level is not usually identical with that at the next level, but it has a strong family relationship. The successive renormalizable approximate theories, by the way, represent autonomous shells that depend on what is inside only through the renormalized parameters.

At the modest level of earlier science, this sort of self-similarity is strikingly apparent. Electricity, gravitation, and magnetism all have the same l/r2 force, and Newton, as we have seen, suggested that there might be some short-range forces as well. Perhaps in some lost manuscript he proposed the Yukawa potential! Now that scientists and mathematicians are paying attention to scaling phenomena, we see in the study of complex systems astonishing power laws extending over many orders of magnitude. Often the underlying mechanism is changing while the power law still holds, as for the cosmic ray energy spectrum, the advance of technologies over time, and so forth.

The renormalization group, which we invented for renormalizable quantum field theory, turns out to apply not only to critical phenomena in condensed matter but to numerous other far-flung subjects as well. The biological and social sciences are just as much involved in these discoveries of scaling behavior as the physical sciences. We are always dealing with Nature consonant and conformable to herself, not only within scaling behavior but also in the occurrence of similar phenomenological laws in a plethora of disparate areas. So the approximate self-similarity of the laws of nature runs the gamut from the simple underlying laws of fundamental physics to the phenomenological laws of the most complex phenomena. No wonder our mathematics keeps working so well in the sciences, when self-similarity is so widespread.

Of course, there may be something important here about the nature of mathematics itself. In connection with that, let me close by paraphrasing some wonderful remarks made by that brilliant and modest theoretical physicist Feza Gürsey on the occasion of his receiving, at the University of Miami, not the valuable kind of prize he deserves but half of the $1,000 Oppenheimer Prize. He said, more or less, that he had achieved some success by pointing, often before other theorists, to mathematical structures that would be useful in the near future in elementary particle physics. But often he hadn't had any clear idea of exactly how or why these mathematical methods would be used. He compared himself with Inspector Clouseau, bumbling along, bumping into walls, but somehow finally pointing to the right suspects. Why, he asked, did the Inspector Clouseau method work? Maybe, he suggested, because such mathematical structures are comparatively rare, so that it is possible to find and identify something like the exceptional group E8 as an object of interest simply because structures with its remarkable properties are not thick on the ground. Thus, it may be that the character of mathematics plays a role in our story, along with Nature consonant and conformable to herself.

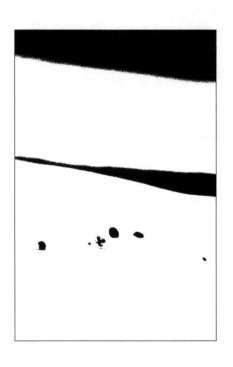

THE SIMPLY COMPLEX: TRENDY BUZZWORD OR EMERGING NEW SCIENCE?

John Casti, University of Vienna
SFI Bulletin, Spring/Summer 1992

A few years ago, I saw a cartoon showing two scientists arguing over the meaning of complexity. In suitably dogmatic terms the first scientist asserted, "Complexity is what you don't understand." Responding to this temerarious claim, his colleague replied, "You don't understand complexity." This circular exchange mirrors perfectly to my eye how the informal term *complexity* has been bandied about in recent years—especially within the normally flinty-eyed community of system scientists—as a characterization of just about everything from aardvarkology to zymurgology. Without benefit of anything even beginning to resemble a definition, we find the putative "science" of complexity being described in terms rosy enough to emit heat: *adaptive* behavior, *chaotic* dynamics, *massively parallel* computation, *self-organization*, and even on to the *creation of life* itself within the cozy confines of a machine. And, to add a final touch of spice, all this hoopla often comes wrapped up in language vague enough to warm the heart of any continental philosopher. But useful as all this fuzziness is for fending off cocktail-party bores and writing research grant proposals, it becomes a major impediment when we start talking seriously about a "science" of complex systems. The problem is that an integral part of transforming complexity (or anything else) into a science involves making that which is fuzzy precise, not the other way around, an exercise we might more compactly express as "formalizing the informal." This short essay represents an exploration into some of the dimensions of this problem, as we try to "scientify" the simply complex.

PHASE I: MAVERICKS

Still More Complex

The science-fiction writer Poul Anderson once remarked, "I have yet to see any problem, however complicated, which, when you looked at it the right way, did not become still more complicated." Substituting the word *complex* for *complicated*, this statement serves admirably to capture the two key points needed to understand what's at issue in turning the casual, everyday notion of a complex system into something resembling an actual science.

The first is the realization that complexity is an inherently subjective concept; what's complex depends upon how you look. So when we speak of something being *complex*, what we're really doing is making use of everyday language to express a feeling or impression that we characterize by the label *complex*. But the meaning of something depends not only upon the language in which it is expressed (i.e., the code), the medium of transmission, and the message, but also upon the context. In short, meaning is bound up in the whole process of communication and doesn't reside in just one or another aspect of it. As a result, the complexity of a political structure, a national economy, or an immune system cannot be regarded as simply a property of that system taken in isolation. Rather, whatever complexity such systems have is a joint property of the system and its interaction with another system, most often an observer and/or controller. So just like truth, beauty, good, and evil, complexity resides as much in the eye of the beholder as it does in the structure and behavior of a system itself.

The second key point brought out by Anderson's quotation is that common usage of the term *complex* is an *informal* one, the word typically employed as a name for something that seems counterintuitive, unpredictable, or just plain hard to pin down. So if it's a genuine science of complex systems we're after, and not just anecdotal accounts based on vague, personal opinions, we're going to have to translate some of these informal notions about the complex and the common into a more formal, stylized language, one in which intuition and meaning can be more or less faithfully captured in symbols and syntax.

These points are pretty obvious, I think, and should hardly be matters of debate among the complex systems crowd. Nevertheless, it's from just such obvious points as these that new sciences emerge by looking at the commonplace and the self-evident in new and interesting ways. And bridging the gap between the informal and the formal is a necessary first step in the making of something that passes for a science out of our intuitive, everyday feelings about the complex. But before entering into a discussion of just how this "subjectivistic" formalization might be carried out, let me pause for a moment to consider why we might want such a thing as a science of complexity in the first place.

Why a Science of Complexity

As noted above, the impression of complexity is something like the expression of an experience of meaning, a part of a cultural cognitive map. And the meaning of our lives depends on the particular maps we use to decode our thoughts, choices, and actions. But human societies have evolved to the point where the traditional maps no longer match our collective experience for very long. Thus, by coming up with a workable (i.e., scientific) theory of complexity, we can hope to be able to internalize the experience of change by describing our collective reality as a process. This, in turn, would then be a major step toward the development of a framework within which we can begin to understand how to control and manage what our maps tell us are complex processes. A second, and somewhat more direct, reason for trying to create a science of the complex is to get a handle on the limits of reductionism as a universal problem-solving approach. When faced with a problem we don't understand, the traditional knee-jerk response is to invoke the old adage "When you don't know what to do, apply what you do

PHASE I: MAVERICKS

know." Most of the time, this translates into an attempt to decompose the "hard" problem into a collection of "simpler" subproblems that we do understand. We then try to reassemble the solutions of these bits and pieces into something that looks like an answer to the original question. Unfortunately, this procedure works just often enough to appeal to the prejudice of reductionists seeking rationalization for their particular brand of epistemological medicine. But we're all familiar with examples like the three-body problem, in which any reductionistic approach of this sort irretrievably destroys the very nature of the problem. Such systems are complex. And it would be nice to have a theory tracing out the boundaries of the reductionistic approach, as opposed to blundering about like blind men, crashing up against these barriers before we even know they exist. So much for motivation. Now let me turn to the twin problems of formalization and "objectification" of the informal and subjective. Let's look at formalization first.

Formalization

The heart of the formalization process is shown in figure 1, which we might term the "modeling relation." Here we see a natural (read:

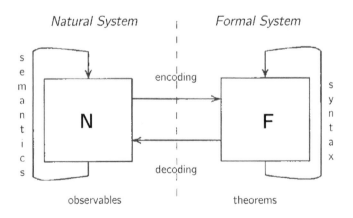

FIGURE 1. The Modeling Relation

real-world) system N characterized by observations and relations stated in everyday language. The formalization process then involves the encoding of these characterizations of N into the symbols and strings of a formal logical (read: mathematical) system F. The key to understanding this process of formalization is to recognize that all notions of meaning (i.e., semantics) reside on the left-hand side of the diagram. So any real-world intuitions we have about N—including its complexity—belong to this side of the modeling relation. By way of contrast, there is no meaning at all on the right-hand side; F consists of mere abstract symbols, together with rules (a grammar) for how strings of these symbols can be combined to form new strings. Whatever meaning might inhere in these strings is then brought out by the decoding of the strings back into N. An example or two will hammer home the point.

The Turing machine

In 1935, Alan Turing was a student at Cambridge University taking a course in mathematical logic. During the course, Turing was exposed to Hilbert's decision problem, and while trying to solve this problem, he invented what we now call a Turing machine. For our purposes here, what's important about this "mathematical computer" is that it represents the first successful attempt to formalize the informal notion of what it means to carry out a "computation." So, despite the fact that people had been calculating for several thousand years, it was not until Turing's work, less than sixty years ago, that the bridge was crossed from the informal system N = computation to its formalization as F = Turing machine.

While Turing's result showing that all computers are created equal is of enormous conceptual significance, I think it's important to point out that very few computer designers, if any, rely upon this fact as they go about their daily chores. So the formalization of the informal idea of a computation has had very little practical impact

PHASE I: MAVERICKS

on problems of modern computer design and operation, despite the fact that the Turing machine serves as the conceptual foundation for a large part of what we now call "computer science."

In passing, let me note another example of the same sort, namely, Gödel's formalization of the informal notion of "truth." Again, not many mathematicians lose any sleep over the implications of Gödel's result for their work. Nevertheless, it's hard to deny the significance of incompleteness as we ponder the soundness of the mathematical enterprise. So it shouldn't come as much of a surprise were we to discover that a successful formalization of complexity will be equally useless from a practical point of view, yet equally profound from the standpoint of setting the foundations for a general theory of models.

Arguing by analogy from the historical genesis of these examples, we see that formalization of the idea of complexity reduces to finding a symbolic structure in F that mirrors our informal ideas about what it is that makes a system complex. In Turing's case, everyday ideas about what it means to carry out a computation were mirrored in the operations of the Turing machine, while Gödel mirrored what we think of as real-world truth by the idea of a mathematical proof. Almost all attempts that I've seen to carry out this kind of mirroring for complexity—information content, length of minimal computer programs, entropy, and thermodynamic depth, to name a few—come down to the translation of informally felt beliefs about the complex into formal, symbolic operations of one sort or another. But so far none of these formal surrogates has achieved a consensus in the system-modeling community as being the "right" formalization.

And despite the numerous interesting and technically deep results that have come out of these attempts, I can't think of a single system modeler whose work is influenced in the slightest by any of these characterizations (which, again arguing by analogy with

Turing and Gödel, could be taken to suggest that they are on the right track after all, I suppose). This sad fact indicates that there's something missing from these formalizations. Recalling the discussion given earlier, let me now argue that the missing ingredient is the explicit recognition that system complexity is a subjective, not an objective, property of an isolated system. But it can become objective, once our formalism takes into account the system with which our target system interacts.

Objectification

Consider a system N and an "observer" who interacts with N. (Here and in what follows, I'll use the emotionally laden term *observer* in the weakest possible sense to mean simply some other system that interacts in some way with N, and not in the strict sense as a system that measures or observes an attribute of N.) The observer creates a linguistic description of the system in the real world. This description is then formalized into a description in the mathematical world F by the process just discussed. We now ask: How many inequivalent descriptions of N can our observer generate? My claim is that the complexity of the system N as seen by the observer is directly proportional to the number of such descriptions. Here's why.

Suppose our system N is a stone on the street. To most of us, this is a pretty simple, almost primitive kind of system. And the reason why we see it as a simple system is that we are capable of interacting with the stone in a very circumscribed number of ways. We can break it, throw it, kick it—and that's about it. Each of these modes of interaction represents a different (i.e., inequivalent) way we have of interacting with the stone. But if we were geologists, then the number of different kinds of interactions available to us would greatly increase. In that case, we could perform various sorts of chemical analyses on the stone, use carbon-dating techniques on it, x-ray it, and so on. So for the geologist our stone becomes a much

PHASE I: MAVERICKS

more complex object as a result of these additional—and inequivalent—modes of interaction. We see from this example that the complexity of the stone is a relative matter, dependent on the nature of the system with which the stone is interacting. And this idea is perfectly general. So how do we get a handle on the number of such "inequivalent" descriptions that are available to a given observer?

Recall that the observer begins by stating an informal description of N in the real world. Then s/he must encode this description into the symbols and strings of a formal logical structure in F. Deciding whether or not two informal, real-world linguistic descriptions are equivalent is a pretty fuzzy affair, opening up all sorts of depressing debates and semantic confusions of the kind that permeate the arts and humanities. But not so in the pristine world of F. Every formal mathematical structure in F comes equipped with its own natural notion of equivalence, a notion that can then be used to classify the informal descriptions. The idea underlying virtually all of these equivalence concepts is that two objects are taken to be equivalent if they can be transformed one to the other by a simple relabeling of the variables used to describe them in F. In short, two objects are equivalent if they differ only in the way we look at them, i.e., by a change of coordinates. So once we have coded our informal description of N into some formal mathematical structure like a set of differential equations, a directed graph, a collection of simplicial complexes, or whatever, the natural equivalence relations for that type of structure can be employed to characterize the level of complexity of the system N. In essence, the complexity level is directly related to the number of equivalence classes that the observer creates by means of the natural equivalence relations defined for the coded version of N in F.

The foregoing idea also provides us with a way to identify when the complexity level shifts as we move through the space of descriptions. Additional complexity appears whenever one description bifurcates from another. So it's exactly the bifurcation points in *F* that can be identified with increased complexity and, even more generally, with emergent phenomena. This observation enables us to relate the complexity of a system to things as seemingly diverse as the eigenvalue structure of matrices, the elementary catastrophes of Thom, and the bifurcation points of vector fields.

A Theory of Models

To summarize, in this essay I've argued that, for complexity to become a science, it's necessary—but far from sufficient—to formalize our intuitive notions about complexity in symbols and syntax. I've further argued that it's necessary for any such formalization to respect the fact that complexity is a subjective concept. One way to do this is to focus attention on the fact that no system lives in isolation. There are always other systems like observers or controllers that are responsible for deciding upon the particular formalization to be used. And, in fact, it is actually these systems that ultimately render the verdict as to what is and isn't complex. So we come to the perhaps not-so-surprising conclusion that the creation of a science of complex systems is really a subtask of the more general, and much more ambitious, program of creating a "theory of models." Complexity—as a science—is merely one of the many rungs on this endless ladder.

LEARNING HOW TO CONTROL COMPLEX SYSTEMS

Seth Lloyd, Massachusetts Institute of Technology
SFI Bulletin, Spring 1995

Scientists and engineers have been hugely successful in solving problems of design and control. Advances in the physical, chemical, biological, and mathematical sciences have been accompanied and driven by the systematic search for technological benefits for society at large. The successes of this search have transformed the ways in which people live and work. One of the most striking transformations of society is the increasing importance of information in providing solutions for problems that were once completely mechanical. For example, a nineteenth-century farmer who wished to provide a cushion against the failure of his wheat crop would plant some fields of corn; today's farmer sells options—bits of information on pieces of paper—to provide a guaranteed income if the crop fails. Where thirty years ago a hot-rodder seeking extra performance would bore out the cylinders of his car, put on dual exhausts and a four-barrel carburetor, a modern hot-rodder simply removes the microprocessor chip that regulates fuel injection and timing, and replaces it with a chip that sacrifices fuel efficiency, low emissions, and reliability for power.

Even when sophisticated information-processing techniques are brought to bear, however, many problems stubbornly resist solution. The initial promises of cybernetics, and, more recently, of artificial intelligence, have proved harder than expected to attain. Some problems in pattern recognition and robotics appear to be simply difficult, in spite of the fact that humans solve such problems every day. These seem to exhibit an intrinsic complexity, a complexity that the process of finding their solutions must share.

PHASE I: MAVERICKS

In addition, though information processing has become so ubiquitous a part of design and control that the humblest of kitchen appliances seems to contain a microchip, success inevitably gives rise to new possibilities for failure. Highly leveraged options traders, when their hedged bets go bad, fail spectacularly. Engines regulated by microprocessors may be efficient and reliable, but they are hard to fix when they break down.

Information Is Physical

From one perspective, dynamical systems can be viewed as simply behaving—obeying the laws of physics. From another perspective, they can be viewed as processing information: how systems get and use information determines how they behave. This article describes research at the Santa Fe Institute and the Massachusetts Institute of Technology that combines these two complementary perspectives to provide a stereo view of dynamics and control. A mathematical framework that treats dynamics and information on an equal footing is used to provide a systematic treatment of how complex adaptive systems face and solve problems of control. This framework, developed by the author with Murray Gell-Mann at SFI, describes how systems get information, how they incorporate that information in models of their surroundings, and how they make decisions on the basis of those models. This framework is designed to support both what Herbert Simon calls blueprints—descriptions of state—and recipes—prescriptions for action—and to provide for their interaction: in our models, adaptation is the coadaptation of blueprints and recipes. Systems learn about their environment by attempting to control it and modify their representation of the environment as a function of the results of those attempts at control. The research taking place at MIT in collaboration with Jean-Jacques Slotine emphasizes specific problems in robotics: how can a robot that has been assigned a particular task, such as catching an irregularly bouncing ball, decide what information is important to gather, how can it best incorporate that information in a model of its task, and how can it learn to perform that task in real time? The answers are relevant to biological systems undergoing natural selection, or to any system that processes information in order to adapt.

History

Historically, though they developed in parallel, dynamical systems theory and the mathematical treatment of information are distinct subjects. After its beginnings in the work of Lyapunov and Poincaré a century ago, dynamical systems theory went through a fifty-year lull before exploding in the last half-century to provide a mathematical understanding first of linear systems, and more recently of nonlinear systems. The methods of dynamical systems theory derive largely from deterministic, classical mechanics.

In contrast, the methods of information theory derive from statistical mechanics, which is probabilistic and, at its base, quantum-mechanical. Information theory originated a century ago when Maxwell, Boltzmann, and Gibbs developed successful statistical treatments of thermodynamics problems. Fifty years later, their work formed the mathematical basis for Shannon's development of a formal theory of information for the purposes of communication, a theory that has expanded to encompass a variety of subjects in computation, control, artificial intelligence, robotics, and a host of other fields.

Although the backgrounds and methods of dynamical systems theory and information theory are historically distinct, in practice, the two fields overlap. This arises from the concurrent development of nonlinear systems theory and of ever-more-powerful digital computers. Computers play a complementary role to analysis in the study of nonlinear systems because the only practical way to follow the individual trajectories of a nonintegrable system is to simulate its dynamics

PHASE I: MAVERICKS

on a computer. That is, in practice, our descriptions of the dynamics of nonlinear systems are at least in part algorithmic—a computational procedure. The method of this research program is to make algorithmic descriptions of dynamical systems a virtue as well as a necessity.

Control and NP-completeness

An algorithm that allows one to simulate the dynamics of a nonlinear system starting from a specific state does not necessarily allow one to control that system. In contrast to the problem of controllability for linear systems, even though following an individual trajectory of a nonlinear system may allow one to verify that a properly chosen sequence of inputs controls the system according to a suitable criterion, it may not necessarily allow one to find such a sequence of controlling inputs or to predict the effect of perturbations on the system and its controls.

Consider the problem of landing an airplane. Suppose that one has programmed a computer to simulate the effect of applying a particular sequence of controls to an airplane in flight. Suppose further that this simulation is both accurate, in the sense that how the airplane actually responds to a particular sequence of controls corresponds closely to the computer's predictions, and efficient, in the sense that the computer can make its predictions of how the airplane responds in "real time," that is, sufficiently rapidly for those predictions to be used to fly the airplane. One can ask, given the assumptions of accuracy and efficiency on the part of the computer, how hard is it to find a set of controls that lands the aircraft safely?

The answer is that, even given the "ideal" qualities of the simulation, the problem of landing the aircraft is still hard to solve. The reason is simple: even though the computer's predictions are accurate, and the effect of a possible sequence of controls can be determined in a short time, there are exponentially large numbers of sequences of possible controls, and to verify the consequences of each one of these sequences in a search for the optimum sequence takes exponential time. That is, if the computer has to answer n yes/no questions properly in order to land the plane, then there are $2n$ different possible sequences of answers, only one of which may actually be the right sequence. Even if it is simple to decide whether or not a given sequence is correct, as n gets large, searching through all possible sequences of yes/no answers to find the right sequence becomes difficult.

A mathematician would say that the problem of trying to find the proper controls to land the plane falls in the computational complexity class NP.

This class consists of problems for which any one potential solution can easily be shown to be correct or incorrect, but for which the space of potential solutions is exponentially large, so that the ability to easily verify potential solutions does not obviously translate into the ability to find a solution. NP problems are in general difficult to solve. If what is meant by an "efficient" simulation of a given system is that the consequences of applying a particular sequence of controls can be evaluated easily, then the problem of finding a sequence of controls that suffices to drive the system to some desired state falls within the computational complexity class NP. That is, even if one assumes that the response of the system that one desires to control can be accurately and efficiently simulated, the problem of finding an adequate set of controls is hard.

In the above treatment, it was assumed that it was possible to predict precisely the result of the application of a particular sequence of controls to the airplane. In real life, of course, whether the plane responds in a desired way to the application of controls is not necessarily fully predictable: For example, the plane may be subject to unpredictable variations in wind speed and shear. Or the plane may respond differently depending on the temperature and density of the air. All such unknown factors can be lumped into a single

PHASE I: MAVERICKS

category called *noise*. Noise represents all that is unpredictable about a problem. A particular set of values of the noise factors is called a *perturbation*, since the noise "perturbs" the system in some random fashion. The presence of noise and perturbations makes the control problem harder.

Usually one does not even know the dynamics of the system that one wants to control well enough to program a computer to simulate them, let alone how to control the system in the presence of perturbations.

In the absence of noise, the control problem can be stated, Does there exist a sequence of controls that drives the system to the desired state? In the presence of noise, the control problem becomes, Does there exist a set of controls that, for all possible unpredictable perturbations, drives the system to the desired state? Adepts in the field of computational complexity will note that where the first problem lies in the computational complexity class NP, the second lies one step farther up the so-called "polynomial hierarchy" of problems: if one is given a sequence of controls, and a sequence of perturbations, then as long as one's simulation is accurate, one can verify easily whether the sequence of controls is adequate to drive the system to the desired state in the presence of those perturbations. But now, even to verify that a given sequence of controls produces the desired state in the presence of all perturbations is itself a difficult problem, in this case in the class Co-NP. Finding an adequate sequence of controls in the presence of noise requires the solution of a potentially infinite number of nested NP problems.

Chapter 6: Learning How to Control Complex Systems

On first sight, these results relating control theory to the theory of computational complexity might seem only to verify in a fancy fashion what engineers have known all along: control of nonlinear systems is difficult. In fact, making explicit the relation of problems of control to NP problems not only shows that these problems are difficult, it shows how one might attempt to solve them. Though seemingly hard in the worst case, many NP problems prove relatively straightforward to solve on average. Over the years, computer scientists and mathematicians have accumulated a library of techniques suitable for attacking various sorts of NP problems. For example, at Caltech, John Doyle has taken a "branch and bound" analysis that has proved successful on a variety of NP problems and applied it to the automatic pilot for the space shuttle to show the existence of previously unsuspected unstable regimes during reentry to the atmosphere.

Adaptive control

Of course, the general control problem is even harder. Usually one does not even know the dynamics of the system that one wants to control well enough to program a computer to simulate them, let alone how to control the system in the presence of perturbations.

To solve problems of control and stability, one needs a picture of the qualitative behavior of the system. That is, for nonlinear systems, control requires insight into the nature of the system's dynamics. Nonlinear control scientists and engineers have developed a number of powerful techniques for characterizing the controllability and stability of nonlinear systems. Some of the most useful of such techniques are Lyapunov's direct method and its extensions, in which the problem of stable control is reduced to finding sequences of inputs that guarantee the uniform convergence of suitably chosen scalar functional to a fixed point. To find a proper such functional in general requires a qualitative understanding of the

PHASE I: MAVERICKS

dynamics of the system in different regimes. For nonlinear systems, control requires intuition.

Traditionally, although techniques such as feedback linearization, optimal control, dynamic programming, etc., have provided a measure of generality in suggesting approaches to control problems, the practical estimation and control of nonlinear systems has proceeded on a system-by-system basis. The necessity for such a case-by-case treatment of nonlinear control problems arises from the lack of robustness for such "universal" techniques when faced with incomplete system identification and perturbations to system dynamics. The goal of this research is to provide a general framework for the systematic construction of algorithms for the control of complex, nonlinear systems in the presence of incomplete system identification and dynamic noise. But, as just noted, control of complex, nonlinear systems requires insight and intuition. How does one provide a framework for the construction of "intuitive" algorithms? An algorithm is a procedure for processing information. To control a nonlinear dynamical system, an algorithm must embody a model for the system: that is, the way in which the algorithm processes information must mirror the way that the system processes information. For the algorithm to model the system successfully, it must be an adaptive algorithm: to acquire intuition, it must learn.

An adaptive algorithm for control alters itself in response to information that it gets about the system that it tries to control. In order to treat the process of adaptation systematically, we introduce a mathematical framework that unifies the description of regular and random features. When represented algorithmically, information about dynamics can be combined with other forms of information to encompass both deterministic and apparently random behavior within the same mathematical framework. The logic behind this unifying framework is simple. The behavior of any system, whether a turbulent fluid, a robot, or the Dow Jones

Chapter 6: Learning How to Control Complex Systems

index, exhibits "regular" features that are predictable and deterministic according to some set of rules—and features that the rules fail to predict—and that are apparently random. Both predictable, regular features and unpredictable, apparently random features can be described using information theory. On the one hand, the amount of information required to specify the regular features can be identified with the length of a suitable computer program that details the set of rules from which the predictable features can be derived, and then derives them. If such a program is of minimal length (i.e., no other program for deriving the predictable features is shorter), then the length of this shortest program is called the algorithmic information of the predictable features. On the other hand, the amount of information required to describe unpredictable, apparently random behavior can be identified with the Shannon information of the ensemble of residual random behaviors of the system after its predictable, rule-based behavior has been specified. These two types of information—algorithmic information to describe rule-based behavior, and Shannon information to describe apparently random behavior—can be added together to give the total information required to describe both predictable and unpredictable behavior.

The idea of combining algorithmic and probabilistic information was recently suggested by [Murray] Gell-Mann and [James] Hartle as a way of making sense of the transition from quantum behavior to classical behavior. They noted that one criterion for identifying the quantum-mechanical operators that correspond to classical systems is to select those that had concise descriptions and whose time evolution was predictable and decoherent. The motivation for introducing total information as a tool for characterizing problems in nonlinear dynamics and control is twofold. First, as noted above, to describe dynamics algorithmically is simply a way of acknowledging and making mathematically precise the importance of computation for characterizing the behavior of nonlinear dynamical systems.

PHASE I: MAVERICKS

Second, and more important, the combining of algorithmic and probabilistic information suggests new methods for addressing problems of measurement, adaptation, and stability that are relevant to the control of complex, nonlinear systems. Information theoretic techniques allow new characterizations of observability and controllability suitable for complex nonlinear systems. In particular, the trade-off between algorithmic and probabilistic information in the formula for total information allows the identification of sampling rates and scales at which such systems are controllable or observable. Of particular interest are what we call "natural" scales and sampling rates: as one goes to finer and finer scales, and to more and more frequent sampling, a scale may arise at which an uncontrollable system suddenly becomes controllable.

According to the particular model that an adaptive controller possesses for the system that it is to control, some of the system's behavior is rule-based, or regular, and some is not rule-based but irregular and apparently random. When the controller adapts, it changes the algorithm that it uses to model the system, and thereby changes what it regards as regular and what it regards as irregular. What the controller treats as order and what it treats as disorder is therefore to some extent arbitrary, and, as the controller adapts, there is trade-off between apparent order and apparent disorder. This trade-off is not a zero-sum game.

Since total information represents the trade-off between rule-based and random behavior, it decreases monotonically as long as addition of extra rules to describe regularities of a system is more than compensated for by a decrease in the system's apparent randomness. This property of total information means that a learning process that minimizes total information is in a sense optimal: arrival at a minimum of total information implies that one has obtained a complete description of the predictable features of a system and expressed this description in the most compact form. In fact, the model for a system's behavior that minimizes total information is optimal in a strict mathematical sense as well: in a maximum likelihood theory of inductive inference, the model that minimizes total information can be shown to be the maximum likelihood model—it is the most concise model that makes the best possible predictions given data. In effect, total information makes mathematically precise the intuition that the world contains both order and disorder, which interact in complicated ways. To characterize and control our surroundings, we must identify the parts of the world where order can be increased at the expense of disorder.

Conclusion

It is no secret that the world is glutted with information. The volume of junk mail, advertising, newsprint, cable channels, PhD theses, and scientific papers is growing exponentially. No matter how quickly they are constructed, the lanes of the information highway soon slow to their customary speed-of-light crawl. In such a world, the capacity to ignore selectively becomes ever more valuable. A system that is to control its environment successfully, be it a consumer, a scientist, or a robot, must adapt by constructing models that allow it to decide what information to get and how to act on it.

BEYOND EXTINCTION: RETHINKING BIODIVERSITY

*Simon Levin, Princeton University, with
Marty Peale, freelance writer
SFI Bulletin, Winter 1995*

In this century, *extinction* has become a household word. We have been alarmed by the data, or numbed by the staggering confirmation that species are going extinct at rates that are four hundred times greater than any recorded throughout geologic time. Researchers sketch the magnitude of the situation in numbers of species lost—tens of thousands annually in the tropics alone—and confirm that the rate of loss is accelerating. Yet even more troubling, as we take stock of the data, we forget how little we know. Indeed, ecologists such as SFI Science Board member Simon Levin, informed by one hundred years of ecological research and, more recently, systems theory, caution that "we barely understand what we are losing."

Individual species, and biological diversity as a whole, have long been valued for aesthetic, ethical, and utilitarian reasons. The most fundamental argument for the preservation of biodiversity is "appreciation of wild creatures and wild places for themselves." The United Nations' World Charter for Nature, for instance, states that "every form of life is unique, warranting respect regardless of its worth to man, and, to accord other organisms such recognition, man must be guided by a moral code of action." The debate about whether to destroy the last remaining strains of the smallpox virus highlights the power of this deep-seated ethic to accord value to all forms of life.

The utilitarian argument for biodiversity rests on the services that are provided to humans. For example, one-fourth of all prescription drugs contain active ingredients originally derived from wild plants. To put this in perspective, the World Wildlife Fund in

PHASE I: MAVERICKS

1991 estimated that only 2 percent of the quarter million described species of vascular plants have been screened for potentially useful chemical compounds. This has prompted great interest in "biological prospecting," the search for potentially useful natural chemicals before they disappear.

Now, in addition to these established arguments, Levin and his colleagues are calling attention to biodiversity as a source of conceptual insights.

A Problem of Definition

We have, at least since Linnean times, recognized the significance of individual species. The definition of *species*, however, is itself problematical. Levin explains,

> Historically species were defined according to morphological characteristics, often on the basis of museum specimens. However, Ernst Mayr and others shifted attention to a biological species concept distinguishing species by the degree to which organisms interbreed. This is a more natural definition, but how do we know that individuals are incapable of interbreeding? Do we mean in captivity or in the wild? What if a species is spatially dispersed, one population breeding successfully with its neighboring populations, but not with distant populations? Where do we split it? And what if a species is temporally dispersed? How do we determine whether a contemporary species is significantly different from its counterpart of 1,000 years ago?
>
> Even "good species" produce fertile hybrids, precisely because there is genetic variation within what we call a species. *Canis lupus* and *Canis canis*, wolves and domestic dogs, are very good examples of this. In difficult cases, researchers may set an arbitrary acceptable level of occurrence of fertile hybrids, such as up to 5 percent, as a

working definition. "Now we're into what a mathematician would call a 'fuzzy set,'" notes Levin, "a continuum, with what we can identify as 'the genetic core of a species'—and gray zones. So a species is not so sharply defined."

Preeminent ecologist E. O. Wilson notes, "Although the species is generally considered to be the 'fundamental unit' for scientific analysis of biodiversity, it is important to recognize that biological diversity is about the variety of living organisms *at all levels*—from genetic variants belonging to the same species, through arrays of species, families and genera, and through population, community, habitat and even ecosystem levels. Biological diversity is, therefore, the 'diversity of life' itself."

"For plant communities," Levin notes, "an ecologist would measure both the within- and among-community components of diversity and recognize each as contributing to diversity. Similarly, in the measurement of biodiversity, one must recognize the diversity within species as well as the diversity in terms of number of species—or, more radically, do away with the notion of species entirely in favor of 'continuum' measures of the genetic and functional diversity of communities."

With these considerations in mind, the significance of species extinctions bears revisiting. Levin asserts, "Species extinction is an important indicator of the loss of biodiversity. It is not, however, the whole story." "Although species extinction is the most fundamental and irreversible manifestation of biodiversity loss, the more profound implication is for ecological functioning and resilience," write fellow researchers Edward Barbier, Jo Burgess, and Carl Folke. Anne H. Ehrlich and Paul R. Ehrlich propose an instructive analogy: "Ecosystems, like well-made airplanes, tend to have redundant subsystems and other 'design' features that permit them to continue functioning after absorbing a certain amount of abuse. A dozen rivets, or a dozen species, might never be missed. On the

PHASE I: MAVERICKS

other hand, the thirteenth rivet popped from a wing flap, or the extinction of a key species involved in the cycling of nitrogen, could lead to a serious accident."[1]

Keystones and Functional Groups

At least since 1935, when Tansley introduced the concept of ecosystems, our understanding of the roles of individual species—and comprehension of the ramifications of loss—have grown. In 1966, Robert Paine introduced the concept of "keystone species," top predators such as starfish and sea otters, whose removal can lead to cascading effects in system properties. Since then, the concept has been extended to species other than top predators. Some, for instance, consider the distemper virus that kills lions in Africa to be a keystone species. Levin cites "a quarter century of research on keystone species—predators, competitors, mutualists, pathogens, among others—demonstrates a diversity of situations in which individual species play critical roles, at least in determining community structure."

Also in the early 1960s, ecologists recognized "keystone groups," or "functional groups." Researchers have since described many systems in which groups of species function as a unit, collectively playing as significant a role as one keystone species plays in another system. Within these functional groups, certain roles are filled interchangeably by one of several species; there is redundancy—"ecosystem insurance," in the words of Stanford University ecologist Harold Mooney. The group as a whole, however, is irreplaceable inasmuch as it controls critical ecosystem processes. One of several *Rhizobium* bacteria, for instance, can fix nitrogen with a legume, the symbiosis thereby enhancing a whole system. Moreover, the

---

[1] Paul R. Ehrlich and Anne H. Ehrlich, *Extinction: The Causes and Consequences of the Disappearance of Species* (New York: Random House, 1981).

diversity within a functional group may be tuned to maintaining resilience to change.

From the viewpoint of systems theory, it is to be expected that large ensembles of interacting components will self-organize into clusters that interact more strongly among themselves than with other such clusters, and that the within-group dynamics will occur on much faster timescales than dynamics among groups. Such hierarchical organization is characteristic of ecosystems.

Robert Steneck and Megan Dethier provide one of the most compelling studies to date confirming the utility of the notion of functional groups. Drawing from experiments and experiences in subtidal algal communities in Maine, Washington, and the Caribbean, they propose that functional groupings of taxonomically distinct species share morphological attributes and that, when the approach of functional groups is taken, convergent biogeographical patterns in ecosystem organization may be discerned clearly.

Communities viewed in terms of functional groupings prove much more stable and predictable than when viewed in terms of species composition. The regularity seen in these communities is reminiscent of the regularity seen in the organization of trophic webs—the network of feeding relationships that defines the pathways of energy flow in an ecosystem—when attention is on the macroscopic properties of those webs rather than on the individual species.

Toward a New Characterization

What, then, is appropriate biotic detail to monitor or preserve? The role of species in ecosystems can be addressed only in the context of how we characterize the boundaries, structure, and function of ecosystems. If the boundaries of the "community" can be stated, for instance, then the shape of the species-area curve is a fundamental aspect of the description of diversity, capturing much more than simply the total number of species in that community.

PHASE I: MAVERICKS

The perspective of the researcher also affects the characterization of the system. A population or community biologist will describe a system in terms of its biotic structure and organization, while an ecosystem scientist will see it in terms of flows and exchanges.

"The health of an ecosystem," notes Levin, "is measured both in terms of its biotic composition and the flow of elements among its compartments." Yet an understanding of the interconnections among these is woefully lacking. "The concept of keystone species gives a handle for understanding community structure; functional groups control critical system properties."

The scientific community barely understands the definition of "individual species," the boundaries of "community," the functional scale at which to characterize "ecosystems," or the interface between "natural selection and self-organization."

The argument for the maintenance of biological diversity is strong in either case. "When species are at issue," explains Levin, "genetic diversity within them governs the capability for resilience. When functional groups are involved, resilience resides in diversity within the groups. Efforts to measure diversity solely in terms of numbers of species, therefore, while a logical place to start, miss much essential detail."

Levin considers HIV to be a particularly instructive example of the functional importance of diversity. "One of the prevailing theories about how HIV operates is that its high mutation rate fosters a diversity that eventually swamps the host's immune system. Do we call that competition between organisms, or cooperation among them to advance 'the genetic core of the species'? Do we define the pathogen as a species, a genus, a functional group? Once again, we are called to question the definition of 'species.'"

How should diversity be measured, and how do system attributes respond to changes in diversity? According to Levin, "We are at the threshold of being able to answer these questions, through combined empirical and theoretical studies. The essential theoretical linkages must come, in part, from a proper theory of ecosystem development and evolution, in which system properties (including organization into functional groupings) are seen to emerge from the self-organizing development of ecosystems and landscapes, within the context of the evolution of individual species."

The scientific community barely understands the definition of "individual species," the boundaries of "community," the functional scale at which to characterize "ecosystems," or the interface between "natural selection and self-organization." In the context of systems and complexity theory, as well as the well-understood aesthetic, utilitarian, and ethical arguments, Levin states, "The idea that humans might make decisions regarding which species to preserve and which to sacrifice is an arrogance that does not sit well on our shoulders." E.O. Wilson has observed that the loss of genetic diversity by the destruction of natural habitats is "the folly our descendants are least likely to forgive us." His prophecy grows only more profound as we probe at the lines we have drawn between things.

WHAT CAN EMERGENCE TELL US ABOUT TODAY'S EASTERN EUROPE?

Cosma Shalizi, SFI and University of Wisconsin
SFI Bulletin, Winter 1999

Eastern Europe is in the midst of a transition of historic dimension. What is the nature of the political, social, and economic arrangements that are forming from the aftermath of communism? Are activities in the countries of Central and Eastern Europe evidence of the "emergent" behavior studied by SFI Science Board member John Holland and others? These questions were the focus of a panel discussion, "Social, Political, and Economic Changes in Central and Eastern Europe," during SFI's recent Fall Symposium, attended by members of SFI's Board of Trustees and Business Network. Loren Jenkins, senior foreign editor at National Public Radio, moderated the event.

John Holland, of the University of Michigan, kicked off the discussion with some general comments about the concept of emergent phenomena, a notion whose definition is still—emerging. Holland probably knows more about emergence than anyone does (that is, as the old joke has it, he's perplexed on a higher and more significant level), but he disclaimed knowing much about Eastern Europe. The other panelists included SFI Trustee Esther Dyson, chairperson of EDventure Holdings, a company focused on new information technology worldwide—particularly the computer markets of Central and Eastern Europe; Harvard University's Anne Goldfeld, a director of the American Refugee Committee; and sociologist David Stark from Columbia University, who has written extensively on emergent economies in the region. Each knew a great deal about Eastern Europe, particularly the final panel member, Lorand Ambrus-Lakatos, an assistant professor of economics and

PHASE I: MAVERICKS

political science at the Central European University in Budapest. However, none were conventional researchers into emergence.

There was, nonetheless, an obvious point of intersection between these two sets of interest and expertise. The countries of Eastern Europe and the former Soviet Union had a decidedly nonemergent, non-self-organized dictatorial socialism imposed upon them, first in 1917 and then in 1945. Starting in the early 1990s, they again were forced into a new form of social organization, what David Stark called a "designer capitalism." Neither form of utopian social engineering met with success. It is scarcely

> There isn't really a theory of emergence, not in the way that physicists have a theory of fluids, or biologists a theory of natural selection. We don't have the right concepts yet.

an exaggeration to say that, to the extent that any part of these economies works, it was not consciously designed, either by the *apparatchiks*[1] of Gosplan (the Soviet Planning Commission) or by the professors from Harvard or other Western institutions, but grew without anyone really guiding it or realizing what they were doing. In many of these countries, for instance, "inter-enterprise" networks have formed as, essentially, a new kind of property relation. Firms bought into each other, daisy-chain fashion, for want of private individuals with the capital to do so. It is just this kind of self-organization that the theory of emergent phenomena is supposed to help us understand. So what can that theory tell us about things like the emergence of new forms of property?

---

[1] Professional functionaries of the Communist Party of the Soviet Union

Chapter 8: What Can Emergence Tell Us About Today's Eastern Europe?

To begin with, it must be confessed, as John Holland did most forthrightly, that there isn't really a theory of emergence, not in the way that physicists have a theory of fluids, or biologists a theory of natural selection. We don't have the right concepts yet. We're not even agreed, all of us, on what counts as emergence, and we certainly can't predict when it will happen, or why, or what form it will take. With respect to what emergence is, we are in the position of the judge who couldn't define obscenity, but knew it when he saw it. Holland says that it has to do with the way agents interact with each other. Moreover, emergent phenomena occur only when the interactions are such that we can't just average the behavior of all the separate individuals to see what they're doing in aggregate. But, as Holland points out in his book *Emergence*, these are necessary but not sufficient conditions. What must be added for sufficiency, for a proper definition of emergence, is yet to be determined.

On the other hand, we do know quite a bit about specific emergent phenomena. One of the lessons of the theory of natural selection (really, of evolutionary game theory) is that a degree of isolation, or "buffering," can be a great help to a new strategy, a new form of behavior, in gaining a foothold. To use David Stark's phrase, "containment can lead to contagion." Take the prisoner's dilemma game, for instance. This is a classic problem of conflict and cooperation in which each of two players has a choice of cooperating with the other or of defecting. If the new behavior is cooperation, it helps greatly if the cooperators can recognize each other and deal preferentially with each other. Then they can even, sometimes, reach a kind of critical mass, at which point others start cooperating out of sheer self-interest. And it may well be that the same thing is true if the game is not prisoner's dilemma but doing business in a newly marketized economy, and the new behavior is not "cooperation" in the senses of the game but fulfilling your end of a contract. There's even some evidence that those countries which, like Poland, privatized

PHASE I: MAVERICKS

and marketized in stages for internal political reasons, rather than submitting to the "shock therapy" of the capitalist international, ended up with more vibrant and successful private sectors; they effectively created such buffered situations—without having any inkling that that was what they were doing. In time, the expectation that people will cooperate, or honor contracts, can emerge as a benign self-fulfilling prophecy, an established and reliable fact of social life—in a word, a convention.

Of course, *emergent* does not necessarily mean *good*. Corruption can become pervasive and conventional in exactly the same way as cooperation, or honest contracting, to the point where even if most people would prefer not to be corrupt, they have to assume that everyone else is, and will take advantage of them if they are not. Which brings us to one of the most important points on which the panel was agreed: markets do not make capitalism. Even capital markets don't make successful capitalism. The fantasy of a frictionless, completely market-driven society is just a fantasy, because successful capitalism depends on nonmarket institutions—schools, police, courts, and all the rest—that are not run along capitalist lines. Also very desirable is a whole vast ecology of social elements that are neither for-profit companies nor parts of the government but that nevertheless comprise "civil society." These, too, can emerge, can form spontaneously. Unfortunately, they haven't in Eastern Europe, at least not as much as they're needed. They may even have shrunk from the days of communism, simply because now so much more effort must go into staying afloat. Clearly, this dynamic is very important but not at all well understood, and it doesn't even seem to have a clear analog among our stock of models of emergence; perhaps mutualism or symbiosis would fit.

At this point, we should introduce two ghosts who haunted the proceedings, the ghosts, aptly enough, of a pair of Central European economists, explorers of emergent and evolutionary phenomena,

Chapter 8: What Can Emergence Tell Us About Today's Eastern Europe?

and of the way market economies fit into the larger society: Joseph Schumpeter and Freidrich Hayek, both originally of Vienna, later of Harvard and Chicago, respectively. They wrote their great works more than half a century ago, and yet echoes of their words could be heard throughout the discussion. Schumpeter's work explains how capitalism requires (and supports) a larger society, many of whose institutions are run on quite antithetical lines. Hayek's explains how markets work as distributed computing mechanisms, adaptively optimizing the allocation of scarce resources, and how society itself is held together by conventions, and the shared expectations they produce. (Admittedly, his work speaks of "spontaneous order" rather than the newer term, *self-organization*.) Today, we have a much better body of abstract theory about emergence, and a wonderful assortment of models, and they make very nice analogies to what Hayek and Schumpeter talked about; Hayek even lived long enough to appreciate some of them. But the question remained, What can they tell us about the real world?

> Of course, *emergent* does not necessarily mean *good*. Corruption can become pervasive and conventional in exactly the same way as cooperation, or honest contracting ... Which brings us to one of the most important points: markets do not make capitalism.

The panel zoomed in from "Eastern Europe" or the "transition economies" to four countries in particular: Hungary, Poland, the Czech Republic, and Russia. A little was said about what used to be Yugoslavia and about Slovakia; other Eastern European countries weren't discussed. The consensus was that Hungary, Poland, and

PHASE I: MAVERICKS

the Czech Republic seem to be on their way to rejoining Europe. Esther Dyson's warnings about the utter lack of role models for the business and investment cultures in the former Soviet Union were troubling, yet Russia may still be able to make progress, with a little luck. But what was happening in Romania, or Georgia, or Kyrgyzstan? Beyond the sobering humanitarian perspective offered by Anne Goldfeld, we heard little, and less that was hopeful. It was implicit in most of the speakers' remarks that success for these countries means looking more like America, or perhaps the slightly imaginary America of an Economics 1 textbook. Everyone worried about the danger of their sinking into the "swamp" of political and economic collapse.

On a good day, we can cobble a common language for sociologists and scientists out of their analogies live on stage. But the real challenge is to come up with something to say to a legislator in Warsaw or an entrepreneur in Kazan or a homemaker in Bucharest that will help them make more sense of their world.

One speaker contemplated a third possibility, that something novel and strange might crawl out of the swamp. It was, David Stark said, possible that some "genuinely new" form of social organization would be produced by the current ferment, something that didn't look much like a liberal capitalist democracy but still, in some fashion, worked. That could be an emergent phenomenon on a very large scale indeed, but it may be a long time before the

Chapter 8: What Can Emergence Tell Us About Today's Eastern Europe?

intellectual descendants of Hayek and Schumpeter and Holland can do anything more to answer such questions than guess, and hope for the best. On a good day, we can cobble a common language for sociologists and scientists out of their analogies live on stage. But the real challenge is to come up with something to say to a legislator in Warsaw or an entrepreneur in Kazan or a homemaker in Bucharest that will help them make more sense of their world.

2000–2014

UNIFIERS

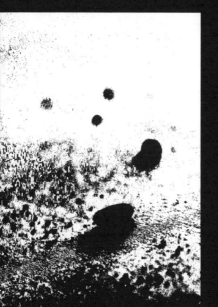

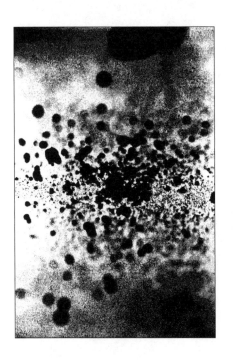

THE EVOLUTIONARY DYNAMICS OF SOCIAL ORGANIZATION IN INSECT SOCIETIES: FROM BEHAVIOR TO GENES AND BACK

Joachim Erber, Technical University of Berlin, and Robert E. Page, Jr., University of California, Davis
SFI Bulletin, Summer 2001

Insect societies have intrigued natural historians and biologists since Aristotle. Scholars have puzzled over the self-sacrificing altruism expressed by sterile colony members—the workers—as well as the complex division of labor and the capability of mass-action responses to the environment. For example, a colony of twenty to thirty thousand honey bees, at one moment in time, may have several thousand individuals engaged in foraging behavior and thousands of others engaged in nest construction, feeding young larvae, or processing honey, while others guard the entrance or thermoregulate the nest. However, when an intruder challenges the entrance of the nest, hundreds or even thousands of worker honeybees may respond immediately by stinging the intruder and, in doing so, sacrifice their lives.

Social insects presented Darwin with major difficulties for his theory of evolution by natural selection. How can you explain the evolution of self-sacrificing worker castes when evolution is a result of the survival and reproduction of individuals? Even more perplexing is the question of how the sterile workers evolved their own traits, different from their reproductive mothers (the queens) when they don't reproduce. But Darwin considered the "acme" of the difficulty to be the presence of distinct anatomical and behavioral castes as seen in many species of ants. How can nonreproductive individuals evolve such a complex caste structure?

PHASE II: UNIFIERS

The social behavior of insects is a result of complex interactions at different levels of biological organization. Genes give rise to proteins and peptides that build the nervous and muscular systems, regulate their own synthesis, interact with each other, and affect the behavior of individuals. Social behavior of an insect colony emerges from the complex interactions of individuals. The interactions that are the fabric of social organization are themselves far removed from the direct effects of the genes, thus providing a major challenge for understanding how insect societies evolve. Understanding how complex societies are organized and evolve is still a central question in evolutionary biology. We now know that, in order to understand how social organization evolves, we must understand the mechanisms that link the different levels of biological and social organization. We must determine the transformational algorithms that link gene to neural system, neural system to individual behavior, and individual behavior to social organization.

This is one of the core themes of the Santa Fe Institute's Program in Evolutionary Dynamics funded by the W. M. Keck Foundation. The aim of the program is to understand how entities with complex organizational structure and function arise and develop; to explore what organizational properties are possible given specific kinds of lower-level components; and to determine the dynamical features that are characteristic of populations of mutating entities that are capable of replication and subject to selection. As self-organized, evolving, complex systems, insect social organizations present perfect case studies for evolutionary dynamics. Last year, former SFI postdoctoral fellow Eric W. Bonabeau and Robert E. Page, Jr. (University of California at Davis), initiated the project "The Evolutionary Dynamics of Social Organization in Insect Societies from Behavior to Genes and Back." Its intellectual ties to SFI's existing research are strong. As mentioned, the work complements other themes within the Keck initiative. Further, the interest and expertise of several Science Board members—Deborah Gordon and Mary Jane West-Eberhard in particular—in evolution, mechanisms, and dynamics of social insect behavior also firmly anchor this initiative at SFI. This effort, now in its second year, draws together people who have common interests and highly complementary domains of expertise, from bioinformatics and genomics to classical animal behavior. The passage from one level of description (for example, genes, neurophysiology, individual insect behavior, colony phenotype) to the next represents an emergent phenomenon of itself and is a research topic of its own. Yet, integrating several such levels of description into an evolutionary model appears to be an absolute necessity if one wishes to have a realistic vision of how evolution operates.

We now know that in order to understand how social organization evolves, we must understand the mechanisms that link the different levels of biological and social organization

---

Social Behavior is Inescapable

In order to understand how social organization evolves, we must understand the mechanisms that link the different levels of biological and social organization. One concept that helps us explore the mechanisms of complex behavioral interactions is the notion of *response threshold*. Neurons and individuals can respond to various stimuli. Responses are based on stimulus thresholds; stimuli below some threshold result in no response while stimuli above a threshold can elicit a reaction. This is evident at the level of neurons where an action potential is generated only when a stimulus sufficiently depolarizes the membrane. Once generated, the action potential

PHASE II: UNIFIERS

propagates at full intensity. At the behavioral level of an individual, we find an analogous process. Individuals do not respond to a stimulus until it is stronger than some minimum threshold. Thus, the response threshold is a fundamental organizing property of the behavior of neurons and individuals.

Because of the stimulus-threshold relationship of behavior, division of labor—the hallmark of social organization—is an inescapable property of group living. This follows because of the correlation between the behavioral response and the effect of the behavior on the stimulus that caused it. For example, honey bee colonies thermoregulate, maintaining the brood chamber of the colony near 34.5°C. When the temperature exceeds this, cooling behavior begins, during which some bees circulate air through the nest by fanning their wings. The result of the response to the heat stimulus, fanning, reduces the stimulus, heat, below the temperature response threshold. This can lead to a division of labor with labor "specialists" when group members have different response thresholds for temperature. Those with lower response thresholds respond first by fanning, reduce the stimulus, and thereby reduce the probability that others will perform that task.

Division of labor self-organizes within groups of cohabiting individuals. This has been demonstrated in two empirical studies. Sakagami and Maeta (1987) forced females of the solitary carpenter bee, *Ceratina flavipes*, to share the same nests. Normally, these bees excavate their own nests by burrowing into the centers of pithy plant stems. After they bore out the center, they forage for pollen and nectar that they bring back to the nest and make into a small breadlike loaf. Then, they lay an egg on the food mass and seal the egg and food in a cell. Several such cells may be constructed in serial order within a single stem. Females guard the entrance of their nest from predators and parasites by effectively plugging the entrance with their own bodies. When pairs of females were forced to share the same nest, a division of labor occurred in every case. One bee became the principal egg layer and guard, while the other did most of the foraging. Hence, a task and reproductive division of labor emerged between these normally solitary individuals. This occurred in the absence of an evolutionary history of nest sharing.

Jennifer Fewell demonstrated a similar phenomenon with the desert harvester ant, *Pogonomyrmex barbatus*. Like most species of ants, young queens excavate a nest in the ground. This nest is sealed and the queen raises her first batch of eggs in isolation, which develop into diminutive workers. The queen catabolizes her own wing muscles to provide the protein needed to raise this initial brood. *P. barbatus* queens normally establish their nests alone. Extensive studies of the species have never found more than one queen constructing a single nest or cohabiting within a nest. However, when females were forced to nest within a confined space, a division of labor spontaneously emerged from their association. In every case, one female did significantly more of the nest excavating. Queens were tested for the amount of time they spent digging before they were paired. In every case, the queen in the pair that did more digging on her own also did more digging when paired. The difference in the amount of digging was amplified by the association, suggesting stimulus-response threshold relationships where the digging stimulus was decreased by the digging activity of the individual with the lowest threshold, reducing the digging activity of the other. Hence, again, a division of labor emerged between individuals without an evolutionary history of cohabitation.

These simple relationships, stimulus-response and the correlation of behavior and the stimulus intensity, represent mechanisms that transform individual behavior into a social organization. The result, division of labor, is an inescapable property of group living because of the behavioral properties of solitary individuals. The evolutionary transition from solitary life history to social life does

PHASE II: UNIFIERS

not require any new genes or new features of the neural system, or new forms of behavior. Insects that live solitary lives already have all the necessary behavioral components for organized social living.

A simple network model can demonstrate how the basic features of insect social organization emerge by depicting the foraging behavior of a colony as an informational network with three parameters: N, the number of elements in the network; K, the "connectedness" of the network (how information is shared); and F, the set of decision functions expressed by the elements. Each element of the network can be "off" or "on." These decisions are made conditional on whether other elements of the network are on or off. Information about the behavior of other elements is made available through network connections. For example, the elements could be foragers and the decision is whether to forage for pollen. An individual that makes the decision to forage for pollen is "on" while one that does not forage for pollen is "off." The forager that is "on" conveys that information to other foragers that then exercise their decision functions. The decision functions are the rules by which they make decisions; for example, "forage for pollen if there are fewer than N other pollen foragers." This would be a threshold-type decision function.

Colonies as Informational Networks

Social insect colonies often consist of a very large number of elements (workers), so it is unlikely that they have an informational net with direct connections between individuals. However, they can share information through sharing common stimuli in the nest. As Robert Jeanne showed for tropical wasps, these stimuli could relate to food stores, young larvae rearing by the colony, nest temperature, or nest construction materials. For example, it has been shown repeatedly that pollen foraging behavioral decisions of honey bees are directly related to the amount of pollen stored in the nest. The amount of stored pollen is a direct function of the consumption of pollen by nurse bees, who convert the pollen proteins into glandular secretions fed to larvae, and the amount of pollen collected by pollen foragers. As more foragers are turned "on" to pollen foraging, there will be more stored pollen, thus providing indirect information to foragers about the pollen foraging activities of others. Division of labor for foraging can emerge from this network of foragers if the activities of the pollen foragers decrease the pollen foraging stimulus. It is likely that many "regulated" activities of social insects are based on similar principles.

Neuroarchitecture and Behavior

The neural systems of insects, consisting of the sensory receptors, interneurons, motoneurons, and synapses connecting neurons, ultimately control the social organization of a colony. The organization of the neural system has many similarities to the organization of the social system. It is composed of many elements, neurons, that are connected in an information network through dendrites, axons, and synapses. The elements are "on" or "off" on the basis of thresholds of response to stimuli in their environment. The elements have information about the current activity states of other connected elements and collectives of elements (neuropil and ganglia) and alter the environment by their activities in ways that affect the response probabilities of other elements.

It seems plausible that the anatomical structure of the neural system itself correlates with behavior. Neuroanatomical studies of Gronenberg and Hölldobler have shown a correlation between visually guided behavior and the size of the eyes in fourteen species of ants. As one would expect, the size of the optic lobes that process visual stimuli correlates with the size of the eyes. In ant workers, large parts of the brain are occupied by the antennal lobes (10 percent of the total brain volume) and the mushroom bodies (20 percent). It

PHASE II: UNIFIERS

is assumed that ants need large brain volumes for olfactory signal processing during the social interactions within a colony. These neuroanatomical studies demonstrate that very general aspects of behavior can be correlated with brain structures. However, we are still very far from understanding how different brain architectures in social insects control different forms of behavior.

Insect learning has many features in common with the learning of higher vertebrates, like us. Learning results in changes in the state of the neural system. Research by Joachim Erber, Uli Müller, and others has shown that uninterrupted neural activity in the antennal lobes and the mushroom bodies are necessary for honey bees to learn odors. The release of modulatory transmitters from specific neurons in response to stimuli initiates a complex cascade of molecular events that ultimately results in cellular changes associated with learning. Specific neurotransmitters and the respective transmitter receptors can induce behavioral plasticity in insects by changing the properties of single neurons and/or neuronal assemblies. We are only beginning to understand the rudiments of the complex neural networks involved in some specific learning conditions. However, it is apparent that there is a self-similarity of organization with respect to learning with common features at the levels of the neuron, neuropil, and the whole animal.

The Evolution of Complex Social Organization

How does division of labor evolve? Selection acting on division of labor takes place at the level of the colony. For the process to occur, there must be heritable genetic variation for social organization. Some colonies survive and reproduce more than others because they have a social organization that is better adapted for a particular environment. Over generations, allelic substitutions take place in populations for genes that affect the behavior of colonial members.

Chapter 9: Social Organization in Insect Societies

There is no "social genome" directing the organization of colonies. The genes responsible for social organization reside in individuals just as they did in their solitary ancestors. In many social insect species, the individuals that demonstrate a division of labor are sterile, the workers. The genes affected by selection must change properties of the neural system that affect the interactions of neurons and the behavior of workers. Changes in division of labor emerge from changes in the interactions of the behavior of the colony members and their environment.

Artificial selection for quantities of pollen stored by honey bee colonies provides an example of how the natural process may work. Nurse bees consume pollen and feed glandular secretions to the larvae. Meanwhile, pollen foragers replace the pollen consumed by nurse bees and maintain a surplus. Rob Page and Greg Hunt selected strains of bees by the amount of surplus pollen they stored in the nest. As a consequence, allelic substitutions occurred for at least three major genes, pln1, pln2, and pln3. These QTLs (quantitative trait loci) affect properties of the neural system that are measured as effects on the response thresholds of bees to sucrose. These effects suggest that allelic substitutions at the QTLs affected signaling cascades in the nervous system (see diagram), but this remains to be

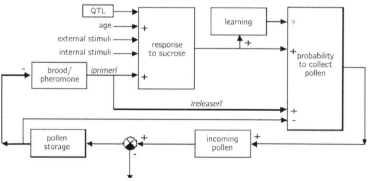

DIAGRAM 1

PHASE II: UNIFIERS

demonstrated. Changes in the nervous system that are expressed as changes in the responsiveness to sucrose clearly affected the probability that a worker will forage for pollen and the acquisition of associative learning, which also may serve to reinforce division of labor. The probability to collect pollen is also affected by stimuli provided by larvae (pheromones produced by the brood) and the quantity of stored pollen. Stored pollen reduces the pollen foraging probability while pheromone produced by the larvae releases pollen-specific foraging behavior. Larval pheromones modulate the probability of foraging for pollen indirectly by changing properties of the neural system that increase responsiveness to sucrose.

Increasing pollen foraging results in more stored pollen and less space for larvae. Colonies use their combs for storing honey and pollen and for raising larvae. The total amount of available space is finite; therefore, if more space is used for storing pollen, less is available for storing honey and brood rearing. As a result, a self-organized negative feedback mechanism emerges between two identified environmental factors that affect the probability to collect pollen. More stored pollen results in a reduction of pollen foraging through its negative effect on the foraging behavior of individuals, but at the same time it reduces the amount of young larvae, thereby reducing the primer and releaser effects of larval pheromones.

The model suggests that small changes at the genetic level may result in very large changes in social organization. Substitution of alleles at a few genes may have profound effects after they are amplified by positive and negative feedback processes at different levels of organization. Perhaps changes in the perception of the environment, such as might occur with small changes in signaling properties of neurons, are the bases of the evolution of at least some functional components of division of labor. A challenge for the future is to better understand behavioral processes at all levels of

Chapter 9: Social Organization in Insect Societies

organization and construct operational algorithms that define the transitions from the genes to the society.

A step toward that end was at the heart of the recent SFI working group gathering titled "Social Insects: Genes, Neurons and Societies," organized by Joachim Erber (Technical University of Berlin), Jennifer Fewell (Arizona State University), and Robert E. Page. The intent of the meeting was to begin to integrate the organizational concepts of neurobiology, behavior, and social biology to develop integrative models between levels of biological organization leading to social behavior and, most importantly, to foster collaborative research projects. The participants formed subgroups focusing on three areas: origins of social organization, division of labor in insect colonies, and communication and signaling. Each subgroup considered a series of overarching common questions and issues, reporting back to the group as a whole. As coordinators, we felt progress beyond expectation was made toward all of the meeting objectives. The success was in large part a result of the organizational structure of the meeting and the composition of the group. We had individuals from very diverse backgrounds in social insects (from genes to ecology) working together in groups determining and discussing the central issues in the evolution of social organization.

The group will reconvene this year to review ongoing individual collaborative projects and to consider the topics of emergent properties of social groups with high and low genetic diversity, theoretic foundations, communication and social integration (regulation of task allocation), and changes in social organization with the number of individuals within groups.

PICASSO AND PERCEPTION: ATTENDING TO THE HIGHER ORDER

Tom Kepler, SFI
SFI Bulletin, Summer 2001

Among the classes I took as an undergraduate physics major, one of the most deeply thought-provoking was Visual Thinking. Art professor Dini Erdman made a significant impact on the course of my scientific career starting with her lectures on figure and ground. First, she taught us the ability to see an inverted relationship between a picture's negative and positive spaces. We can easily experience these shifts in the representations developed by Edgar Rubin and exemplified by his faces/vase picture. One's sense of the object of perception shifts dramatically as the figure–ground relationship flips.

More to the heart of the aesthetic experience is the construction of visual art based on the dynamics emerging in the tension between these two complementary spaces. The Picasso painting *Les Demoiselles D'Avignon* is a particularly striking example of this technique. The "empty" space between the women and between their arms and torsos, for example, is depicted with a certain solidity of its own. This does not simply invert the figure–ground relationship but forces the viewer to abandon the traditional way of parsing a scene and instead attend primarily to the higher-order perception arising in the relative relationship between the fields rather than to either field absolutely.

I know that for actual visual artists, this is preschool stuff, but somehow it seemed a revelation—the difference, perhaps, between having been taught and seeing directly.

This was my first lesson in complex systems.

And I think it still paints an accurate, if simplified and poetically suggestive, picture of what goes on at SFI. We frequently talk

PHASE II: UNIFIERS

about seeing the "big picture," but, of course, merely seeing the big picture is insufficient; seeing the big picture necessitates the loss of information at higher resolution. What goes on at SFI is not simply a withdrawal to a vantage point providing a more encompassing vision. Rather, it's more akin to the development of a new way of directing attention and deploying it toward seeing and manipulating the tension between what is currently foreground and what is background. Agent-based modeling itself provides a prime example of this. One's attention necessarily shifts between the detailed behaviors of the agents themselves and the emergent collective phenomena they embody until a kind of synthesis or merging of the two perspectives occurs and there arises a new locus of attention.

And this is reflected also in the transdisciplinary strategy. In any given discipline, a concerted effort is made to train a bright light on a particular set of issues for which theoretical and/or experimental tools have been painstakingly crafted. A special language is developed that facilitates communication about these foreground matters, which inevitably casts further shadows on those topics implicitly relegated to the background.

When practitioners of different disciplines meet to share ideas and work collaboratively, this difference in language is immediately clear and frequently troublesome. The language of physics does not lend itself easily to discussions about the economy; nor does the language of economics lend itself to physics. The task of mutually interpreting these languages requires a partial inversion of the foreground–background distinction, at least temporarily, on both sides. At this point, new parts of the landscape are illuminated, though largely by the light of the old lamps. When these collaborations are successful, a new metalanguage eventually emerges. But here, too, this metalanguage does not correspond to a new set of foreground objects so much as to a higher-order set of relationships at the boundaries between figure and ground.

The International Program, now in its second year, may be pointing beyond mere interdisciplinarity. The stated purpose of the International Program is to disseminate the ideas and methods underlying the analysis of complex systems in the developing world. We have now begun to build networks of interested researchers in China, India, Africa, the former Soviet Union, and Eastern and Central Europe, among other places. The first couple of workshops have now happened as well as many more informal meetings. What has emerged in these gatherings is the possibility of taking the methods of transdisciplinary research to a new level.

In contrast to the present practice of such research at SFI, where all parties' disciplinary languages are embedded in common intellectual tradition and experience, we now see a different type of interaction. The researchers may be talking about the same systems but doing so from different points of view. One particularly clear example of this potentially powerful transcultural collaboration is in medicine, where the languages of Asian traditional medicines are so orthogonal to that of Western biomedicine that it's hard to accept that both, simultaneously, have legitimacy. But recent metastudies, including one from the National Institutes of Health, have shown just that. I suggest that here, too, we have a figure–ground reversal, only this time the participants are all looking at the same thing. But where biomedicine sees substance, Eastern traditions see process.

There are similar conditions, though perhaps not as starkly defined, that inform the dialogue between the pragmatic scientists of the US and the more philosophically motivated researchers of Eastern and Central Europe: same objects, different languages. Shall we resist the temptation to simply translate and instead seize upon the opportunity to shift our gaze to embrace both representations simultaneously?

In college I also learned that one cannot cover a sphere with a single smooth coordinate system. Instead, one must devise two smooth coordinate systems and the rules for connecting them where they overlap. A single universal and complete representation simply cannot cover spaces with nontrivial topology. Shall we continue to bet that a single universal and complete representation covering any given complex system—much less the world—is to be found?

FOUR COMPLICATIONS IN UNDERSTANDING THE EVOLUTIONARY PROCESS

Richard C. Lewontin, Harvard University
SFI Bulletin, Winter 2003

Taxonomic Space

In order to discuss complications that arise in the understanding of evolutionary processes, it is first necessary to make clear what the evolutionary explanation is to accomplish. For this purpose the concept of *taxonomic space* is a useful one. We owe this notion to G. Evelyn Hutchinson, but Walter Fontana and others have since used it in one form or another. This taxonomic space of organisms has a huge number of dimensions, each corresponding to some character that might be used in the characterization of an individual. If one looks at the occupancy of such a space, one is struck by the fact that it has a structure to it. Individual organisms are clustered in the space and those clusters are themselves clustered. And there are clusters of clusters of clusters, rather like the stars in the cosmos. The most important thing for the evolutionist is that nearly the entire space is empty, not only when extant organisms are considered, but when all organisms known to have ever existed are considered. The measure of the emptiness of that space is nearly one, and the measure of the occupancy is nearly zero.

The real problem for the evolutionist is not to explain the kinds of organisms that have actually ever existed. The real problem for the evolutionist is how it is that most kinds of potential and seemingly reasonable organisms have never existed. The problem is to explain the location of the empty spaces in the clustered assemblage of occupied points. It is easy to describe organisms that have never

PHASE II: UNIFIERS

existed. There are snakes that live in the grass, but there are no grass-eating snakes. Birds perch in trees, yet, aside from a few exceptions, they do not eat all that greenery around them but rather spend a great deal of energy searching for food. So why are there virtually no leaf-eating birds? The fact that the measure of the unoccupied space is so big compared to the measure of the occupied space means that explanations of that lack of occupancy are not so easy to come by. That most of the space is empty is expected since the dimensionality is enormous and only a relatively small number of organisms have come into existence since the beginning of life. Since there has only been one history of life, the reason for the low occupancy in the total space is the finiteness of time.

> We will never evolve into a race of angels because we do not have the genotype that will allow for the possession of arms, legs, and wings.

Hierarchical Clumping

The structure of the occupancy is another matter. Organisms are underdispersed in taxonomic space, and we need to understand the causes of the hierarchical clumping. One reason for hierarchical clumping in taxonomic space is simply that organisms arise one from another. If an organism is someplace in taxonomic space, it is likely that its immediate descendants will be someplace close by in the space rather than someplace far away. It may not be that a particular region in the space is impossible to fill or that you can't get there from here, but that there has not been enough time for evolution to fill that space.

On the other hand, the structure of accessibility may make it impossible to get there from here without retracing the steps to a remote branch point that led from a distant ancestral state. One remarkable evolutionary example of not being able to get there from here is that no vertebrate has ever succeeded in evolving wings without giving up something. There are no hexapod vertebrates. Bats and birds have had to give up their forelimbs to produce wings. We will never evolve into a race of angels because we do not have the genotype that will allow for the possession of arms, legs, and wings. There is no general structural problem of evolving multiple limbs and multiple wings. Insects have succeeded in evolving six legs and four wings. So the problem for vertebrates is that of not being able to get there from here without retracing the evolution of vertebrates from invertebrates. In the absence of a very large number of trials such as we have in the case of the entire collection of vertebrates, we cannot know whether a specific "hole" in the space is a consequence of the structure of accessibility or simply the chance result of a small sample size.

Taxonomic space may be clumped because there are ways of making a living that are so costly or have such a low survivorship and competitive ability in the face of already-existing organisms that natural selection has prevented their occurrence except as rare mutational forms. Finally, there are some processes and structures that may simply not work given the general structure of the organisms in which they might occur. Despite the immense variation in methods of locomotion that animals have evolved, there are no organisms that move along the ground on wheels. Presumably this is a consequence of the problem of enervating and supplying nutrients to an axially rotating macroscopic structure.

Four Complications

When we contend with "innovation" and "novelty" in evolution, we are concerned with the occupation of a region of the taxonomic space that has been previously empty. Unlike Walter Fontana's

PHASE II: UNIFIERS

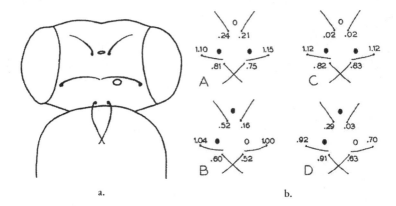

FIGURE 1. Ocelli and ocellar bristles in *Drosophila*. a) Normal pattern; b) result of selection for two ocelli in lines selected only for posterior ocelli (A,C) and for one anterior and one posterior ocellus (B,D). Numbers indicate the mean number of ocellar bristles at each position. (From Maynard Smith and Sondhi, 1960).

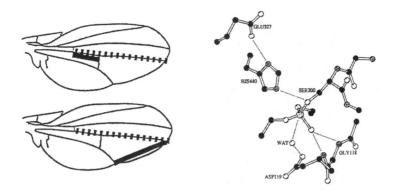

FIGURE 2. Wing dimensions measured in the experiment of Haynes (1989). Dashed line is the control length. Solid lines are the lengths whose ratio to the control length is used as a basis for selection.

FIGURE 3. Molecular structure of active site of cholinesterase in *Lucilia coprina*, showing the water molecule bound at the aspartic acid residue (119) and the phosphate bound to the serine residue (200). (From Newcomb et al., 1997).

usage, novelty for the biologist is not the occupancy of a state that is somehow "difficult" to get to but rather the more intuitive notion of the occupancy of a state that is a surprise, because it has never happened before despite a very large number of trials. Such novelties need not be very distant in the space from already-existing forms, and they need not be very large unoccupied regions, but may be in taxonomically quite small subspaces, as for example the evolution of a grass-eating snake. It is the pathways of evolution of novelties as I have defined them here that have shown a variety of possibilities that are themselves unexpected and whose occurrence should caution us against making easy model assumptions about what it takes to produce an evolutionary novelty.

Empty space doesn't tell us much

The first thing that we must take into account is that we cannot judge how easy it is to create a novelty from the simple observation that parts of taxonomic space seem to have been avoided by organisms. There is a vast literature produced during the middle of the twentieth century showing that there exists within species a large reservoir of standing genetic variation that can be used by selection to move a population to a position in the space that is not only currently unoccupied but appears to be prohibited by some genetic or developmental constraint. The best-known cases are those in which some aspect of the phenotype is invariant within a species, but variation between individuals can be produced by stressing the development either genetically or environmentally. By selecting among the variants, the mean phenotype of the population can be changed and this change is heritable, demonstrating that there was genetic variation relevant to the character in the population but that under normal developmental circumstances this variation was hidden. That is, the development of the phenotype was buffered or "canalized" (Waddington 1953; Rendel 1959). Such changes can

PHASE II: UNIFIERS

alter a character that is invariant not only within a species but over a large taxonomic range, as for example placement of the three simple light receptors (ocelli) and their six associated sensory bristles on the head of all individuals in all species of the genus *Drosophila*. The ocelli are normally symmetrically placed, one anterior to the left, one to the right, and one posterior on the midline of the head (fig. 1a). Maynard Smith and Sondhi (1960) succeeded in creating lines with only the posterior ocelli and, more surprisingly, lines in which the majority of individuals were asymmetrical (fig. 1b).

What is less well known is that allometric shape patterns that appear to be the consequence of unbreakable allometric relations that apply over many species of different size can also be changed by genetic variation already present within species. An example is the experiment of Anna Haynes (1989) on wing dimensions in *Drosophila*. Figure 2 shows two wing vein lengths that are negatively correlated among individuals within all species of *Drosophila* and between species means of all species in the genus. Haynes selected individuals in *Drosophila melanogaster* for which both measurements were larger (relative to a control length on the same wing) than the mean and, in another selected line, in which both were smaller than the mean. As a result, in only fifteen generations she succeeded in changing the correlation between the measurements from $-.4$ to $+.2$, breaking a genus-wide correlation. Such a genus-wide correlation seems an obvious candidate for a basic developmental constraint, yet the experiment shows that it is trivially easy to break using the genetic variation that is already present in the species.

In this case, we must conclude that the unoccupied region of the phenotypic space is easily accessible genetically and developmentally but is empty because of natural selection. The same phenomenon was demonstrated for anterior and posterior eye spots on the wings of the butterfly *Bicyclus anynana* by Beldade et al. (2002). A strong positive correlation in the size of anterior and posterior eye spot size and other serially repeated features is the rule in butterflies and has been assumed to be a consequence of basic developmental mechanisms of anterio-posterior differentiation. The experiment reversed the correlation within eleven generations of selection. In both cases, despite the universality of the correlations in nature, there was enough genetic variation in growth relations within a population to allow a selective reversal within a few generations of the pattern.

A thorough aerodynamic modeling of the relation between fly size, lift, and wing dimensions in *Drosophila* might reveal a functional rule for the case of the fruit flies. But there are other selective reasons besides immediate function that keep regions of the space empty. There is a large literature showing that *Drosophila* females discriminate in their acceptance of courting males against individuals who deviate from the usual morphology for the species, as for example, deviant eye or body colors. It is this discrimination that prevents mating between species, but it also keeps the morphology of a given species within narrow bounds. It is part of the theoretical commitment of "evo-devo," the study of the evolution of development and the influence of developmental pathways on evolution, that shape is greatly constrained by basic developmental relations resulting from cell-to-cell signaling and gradients in gene transcription that are more or less fixed across a wide range of organisms. That may indeed be true for some features of development, but it is also clear that the observed constancy of some feature is not in itself a demonstration of such genetically determined invariance. At least for wings in flies and moths, we must assume that natural selection is playing a stabilizing role in preventing evolutionary change in these organisms that is already possible with the genetic variability that they possess.

PHASE II: UNIFIERS

Small changes lead to functional novelties

The second complication is that what we judge to be extremely small changes can produce what everyone would agree to be functional novelties. An example is a case in which a biochemical novelty may arise by a single very small molecular change. Newcomb, Campbell et al. (1997) found that the acquisition of organophosphate herbicide resistance in the blowfly, *Lucilia coprina*, is a consequence of a single amino acid substitution in the active site of a carboxylesterase that abolished that enzyme specificity and converted the enzyme to an organophosphatase. Figure 3 shows the three-dimensional structure of a closely related esterase with essentially the same structure as the carboxlyesterase at the active site. The amino acid mutation that changed the function was the substitution of an asparagine residue for a glycine that allows a water molecule to be bound near the site of binding of the organophosphate. The structural change allows the molecule to participate in an attack on the phosphate bond, hydrolyzing it and destroying a molecule of the organophosphate. Thus, the qualitative change in specificity was a consequence of a small change in the angle at which the substituted amino acid was held in the folded molecule. That this change was not an extraordinary event was shown by the discovery of a second, different amino acid substitution that had the same effect. So, small genetic changes may lead to novel adaptive consequences.

Getting there from here

A third complication in the process of evolutionary change arises from the topology of accessibility of states, the problem of "getting there from here." One of the most illuminating and well-understood cases at the genetic level is Barry Hall's selection of a novel biochemical function in *E. coli*.

Hall (1978) set about to select *E. coli* that could use a novel carbon source, lactobionate, for its energy, instead of the usual lactose. For this purpose he used a gene, *ebg* (extra beta galactosidase), that had a low efficiency for cleaving the galactosidic bond of lactose and could be dispensed with in normal lactose metabolism. The first step in the experiment was to knock out the *lac* gene that codes for the normal beta-galactosidase, making a strain that required the *ebg* gene for normal lactose metabolism. Using a mutagen, he succeeded in accumulating mutations of *ebg* that would allow growth on lactobionate, but the evolutionary path to this state was not direct. He was not able to select directly for the new substrate. First, he had to select for a control mutation such that the *ebg* gene would be transcribed even in the absence of lactose as an inducer of transcription. Next, he had to select for increased activity on lactose. Then these first selected stages had to be followed by a stage of selection for an intermediate substrate, lactulose, and then a strain that could ferment lactulose was successfully selected to grow on lactobionate. Moreover, at each stage there were several strains that possessed the same biochemical phenotype, but only some of them could be further selected to the next stage. This result illustrates that the pathway through the space of genotypes from one phenotypic state to another is complex, rather like a maze with many dead ends. Only a restricted subset of all the pathways that lead to the first adaptation are open to the next so that evolution of a novelty may be very difficult to achieve. This suggests one reason for the apparent conservatism of intermediary metabolism.

Differential fitness

Finally, we must consider the way in which differential fitness constrains the occupancy of the taxonomic space. Unfortunately, the determination of fitness is a great deal more complicated than is usually supposed. It is easy to say that fitness of a type is its "relative probability of survival and reproduction," but turning that phrase

PHASE II: UNIFIERS

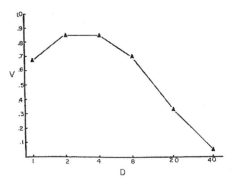

FIGURE 4. Survival of *Drosophila* larvae as a function of density. (From Lewontin, 1955).

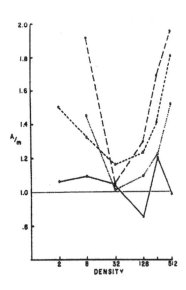

FIGURE 5. Relative survival of two competing larval genotypes of *Drosophila busckii* as a function of density and composition. Solid line: predicted relative survival from pure culture experiments of Acme and mutant. Long dashes: observed relative survival in 75% Acme cultures; Short dashes: observed relative survival in 50% Acme cultures; Dots: observed relative survival in 25% Acme cultures. (From Lewontin and Matsuo, 1963).

into a coherent measure that can do work in evolutionary explanation is not so easy.

First, it is obvious that the fitness of a type depends on the environment in which the organism lives. But the environment is not independent of the organism. Organisms, by their biology, determine what aspects of the external world are relevant to them and constantly change their environment by their life activities. That means that as a collection of organisms evolves, their environment evolves with them. The evolution of organism and environment may be described by a pair of coupled differential equations in which changes in both organism [d(org)] and environment [d(env)] are functions of both variables:

$$d(\text{org})/dt = f(\text{org}, \text{env})$$
$$d(\text{env})/dt = g(\text{org}, \text{env})$$

A consequence of the codependence of the properties of organisms and their environment is that the Darwinian fitness relations among competing types can be very complex. In particular, the relative fitnesses of genotypes may depend both on the population density of the organisms and on the relative frequency and identity of the competing types. An example of this can be seen in experiments on the effect of population density and composition in *Drosophila* (Lewontin 1955; Lewontin and Matsuo 1963). In these experiments newly hatched *Drosophila* larvae were placed on a measured amount of an agar medium on which yeast was seeded. An example of a typical result is shown in figure 4 for an experiment on *Drosophila melanogaster* where the absolute probability of survival to adulthood of different genotypes was measured at different population densities. The highest probability of survival is not at the lowest density but at an intermediate density (four to eight per vial). This intermediate optimum is a consequence of the larvae tunneling in the agar, which increases the surface area for yeast growth that is

PHASE II: UNIFIERS

the food of the larvae. The effect can be abolished by making the food so soft that no tunnels are produced. The next step is to mix larvae of different genotypes at various densities to observe the relative probabilities of survival in competition. A typical result is shown in figure 5 from an experiment on *Drosophila busckii*. The solid line is the predicted relative survival of two genotypes at different densities, the prediction coming from the absolute survival of the genotypes in pure culture. The dashed and dotted lines are the observed relative survivals in mixed culture at the various densities. What figure 5 shows is that only at the optimal density (thirty-two per vial for this species) is the actual relative survival predictable from the pure culture survivals. At the nonoptimal densities, one genotype is superior to the other, and the degree of this superiority depends both on total density and on the relative proportion of the two genotypes. That is, the force of selection is both density and frequency dependent.

In modern evolutionary theory, "fitness" is no longer a characterization of the relation of the organism to the environment that leads to reproductive consequences but is meant to be a quantitative expression of the differential reproductive schedules themselves. Darwin's sense of *fit* has been completely bypassed.

The complications that arise from frequency dependence are even greater than those shown in the previous experiment. In experiments involving competition of several genotypes taken two at a time, Dobzhansky (1948) showed lack of transitivity of fitness. That is, genotype A is more fit than genotype B in an experiment involving only these two genotypes, and B is more fit than C in two-way competition, but in three-way competition C beats A. If organisms play a game of scissors–paper–stone in which there is no simple transitivity of differential fitness, then no prediction of the actual outcome or application of game theory that depends on standard utility theory is possible without a detailed mapping of the fitness or utility space.

The difficulties of the concept of fitness are, unfortunately, much deeper than the problem of frequency and density dependence. The problem is that it is not entirely clear what fitness is. Darwin took the metaphorical sense of fitness literally. The natural properties of different types resulted in their differential "fit" into the environment in which they lived. The better the fit to the environment, the more likely they were to survive and the greater their rate of reproduction. This differential rate of reproduction would then result in a change of abundance of the different types.

In modern evolutionary theory, however, "fitness" is no longer a characterization of the relation of the organism to the environment that leads to reproductive consequences but is meant to be a quantitative expression of the differential reproductive schedules themselves. Darwin's sense of *fit* has been completely bypassed. The natural properties of organisms lead to differential reproductive schedules, and these must somehow be mappable onto a quantitative function, fitness that can enter into formal prediction structures. There is also an implication that fitness is a scalar quantity since much of the informal argument of evolutionary theory characterizes one type as "more fit" than another. To make such a scalar work in prediction, a standard viability model of reproduction has been created in which the organisms have discrete generations so that all can

PHASE II: UNIFIERS

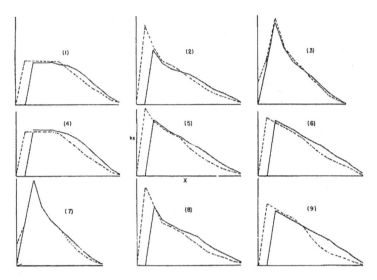

FIGURE 6 Pairs of contrasting kx schedules for which frequency changes were calculated in populations of increasing and decreasing size. Abscissa: kx, ordinate: age, x. (From Charlesworth and Giesel, 1972).

be regarded as being born simultaneously and all differences in fitness are the different probabilities of survivorship to sexual maturity.

Any relaxation from the standard viability model produces serious problems in the definition of fitness. If there are differences in fertility and the organisms are sexually reproducing, then fertility, in the general case, is a function of the mating pair. Averaging over different mating combinations will provide a mean fertility of each genotype, but such means are necessarily frequency dependent, so the quantitative values will change during the evolutionary process and even the ordering of type fitnesses may shift. The fitness of a genotype can then not be assigned apart from a statement of its frequency in the population and the rules of mating preferences. If we further relax the standard viability model to include all those species with overlapping generations and reproduction that occur over an extended period of the individual's lifetime, then the totality of the reproductive information consists in the age schedule of relative mortality and fertility of different types, embodied in the probability of living from birth to age x, l_x, and the number of offspring, b_x, produced by an individual of age x in the interval x to x + dx. If the species is sexually reproducing, the vector of age-specific fertilities bx must be substituted by a matrix of the fertilities of couples B_{xy} of females aged x and males aged y for each genotypic composition of the pair. These are then averaged to produce a matrix of frequency-dependent means for each genotype. These values change not only as frequencies change but as the population changes its age distribution. The attempt by Fisher to circumvent these complications by defining the fitness of a genotype as the root m of the Euler equation did not solve the problem because it confuses the rate of reproduction of a type with the rate of reproduction by a type, which are not at all the same thing in a sexually reproducing species, and also assumes that the population is at the stable age distribution, which is not true for a population changing its type frequencies. But the problem is even worse.

It is the case that all the information about the relative reproductive behavior of types in the population is contained in the complete l_x and b_x schedules of all the genotypes (and, for sexually reproducing species, the age schedule of mating pairs and the frequencies of the different types). Yet this complete reproductive information is insufficient to predict whether a type will increase or decrease in frequency in the population! It is also necessary to know whether the population as a whole is growing larger, is stable in numbers, or is decreasing in numbers. The same type that may be favored in a growing population may be disfavored in a shrinking population. Suppose the only difference between two types is not in their total reproduction but in their age schedule of progeny production. A type that produces offspring at an early age will

PHASE II: UNIFIERS

increase in relative frequency in a growing population because it has reproduced while the total population is still small. If the population is shrinking, however, it pays to postpone reproduction since the total population will then be smaller at the time of reproduction of the tardy type.

Unfortunately, a simple examination of the reproductive schedules does not always reveal that one schedule is obviously "back-loaded" and one "front-loaded," as economists would put it. Figure 6 from the work of Charlesworth and Giesel (1972) shows a number of pairs of hypothetical relative reproductive schedules expressed as k_x, the product of l_x and b_x. In cases 4, 5, 6, and 7, which of the two schedules was favored depended on whether the population was increasing or decreasing in total size. In cases 1, 2, 3, 8, and 9, there was no such contingency. There is no obvious common feature that would have allowed us to predict these classes. How, then, are we to assign relative fitnesses of types based solely on their properties of reproduction? But if we cannot do that, what does it mean to say that a type with one set of natural properties is more reproductively fit than another? This problem has led some theorists to equate fitness with outcome. If a type increases in a population, then it is, by definition, more fit. But this suffers from two difficulties. First, it does not distinguish random changes in frequencies in finite populations from changes that are a consequence of different biological properties. Finally, it destroys any use of differential fitness as an explanation of change. It simply affirms that types change in frequency. But we already knew that.

REFERENCES CITED

Beldade, P., K. Koops, and P. M. Brakefield. 2002. "Developmental Constraints Versus Flexibility in Morphological Evolution." *Nature* 416: 844–47.

Charlesworth, B., and J. T. Giesel. 1972. "Selection in Populations with Overlapping Generations, II. Relations Between Gene Frequency and Demographic Variables." *American Naturalist* 106: 388–401.

Dobzhansky, T. 1948. "Genetics of Natural Populations, XVIII. Experiments on Chromosomes of *D. Pseudoobscura* from Different Geographical Regions." *Genetics* 33: 588–602.

Hall, B. 1978. "Experimental Evolution of a New Enzymatic Function, II. Evolution of Multiple Functions for EBG Enzyme in *E. Coli*." *Genetics* 89: 453–65.

Haynes, A. 1989. "On Developmental Constraints in the *Drosophila* Wing." PhD thesis. Harvard University, Cambridge, MA. 115 pp.

Lewontin, R. C. 1955. "The Effects of Population Density and Composition on Viability in *Drosophila Melanogaster*." *Evolution* 9: 27–41.

Lewontin, R. C., and Y. Matsuo. 1963. "Interaction of Genotypes Determining Viability in Drosophila Busckii." *Proceedings of the National Academy of Sciences of the USA* 49: 270–78.

Maynard Smith, J., and K. C. Sondhi. 1960. "The Genetics of a Pattern." *Genetics* 45: 1039–50.

Newcomb, R. D., P. M. Campbell, D. L. Ollis, E. Cheah, R. J. Russell, and J. G. Oakeshott. 1997. "A Single Amino Acid Substitution Converts a Carboxylesterase to an Organophosphorus Hydrolase and Confers Insecticide Resistance on a Blowfly." *Proceedings of the National Academy of Sciences of the USA* 94: 7464–68.

Rendel, J. M. 1959. "Canalization of the Scute Phenotype of *Drosophila*." *Evolution* 13: 425–39.

Waddington, C. H. 1953. "Genetic Assimilation of an Acquired Character." *Evolution* 7: 118–26.

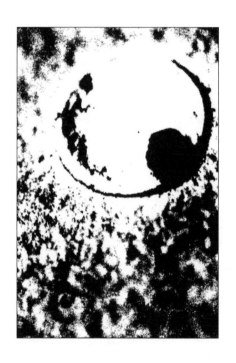

SEARCHING FOR THE LAWS OF LIFE: SEPARATING CHANCE FROM NECESSITY

*D. Eric Smith, SFI, and
Harold J. Morowitz, George Mason University
SFI Bulletin, Winter 2004*

The noted contemporary paleontologist and natural historian Steven Jay Gould has said of the history of life that "any replay of the tape would lead evolution down a pathway radically different from the road actually taken" (Gould 1989, 50). Should one make such a strong statement about all aspects of life, though? Gould studied the body plans of the major groups of animals that suddenly appeared in the fossil record 570 million years ago, in a period called the Cambrian explosion. Indeed, it seems largely accidental that just this combination should have come to make up the entire animal world, creating a large-scale taxonomy of which only a sub-part has survived to this day.

But what about the chemical composition of those organisms, or the way they capture energy to maintain and replace themselves, which we also share? Could that really have taken a different form than the one we see attested today? What about the great events when biological innovations changed the surface chemistry of the earth, like the emergence of photosynthesis that loaded our atmosphere with molecular oxygen, after two billion years in which it had had very little? What of endosymbiosis, when one group of bacteria-like unicells began living as organelles within another? How much of chance is there in these stages of our shared structure and history, and how much of necessity?

For five weeks in the summer of 2003, a diverse group led by Science Board member Harold Morowitz, postdoctoral fellow

PHASE II: UNIFIERS

Jennifer Dunne, and research professor D. Eric Smith met to examine some of the universal structures and patterns in living systems, from biochemistry to ecology, and to ask which might have arisen from the action of underlying "laws of life." The goal was a set of rules or principles that select living forms from chemistry and geophysics, the way simple rules such as the Pauli exclusion principle generate the periodic table of the elements, and all of chemistry, from a few properties of the proton, neutron, and electron.

> The beautiful weblike cell wall that encrusts all bacteria like a Fabergé egg or the ubiquitous scaling laws in ecology.

The discussion ranged from narrow technical details of core biochemistry to broad philosophical questions of what should be meant by "laws" in biology. It is clear that, while biology is a natural science whose observations can be quite precise and often quantitative, the biological notion of *understanding* assigns less importance to predictions about the specific course of the future than is given in chemistry or physics. The roles of accident, individuality, and uniqueness are correspondingly greater in biology, and with these it becomes less clear how to interpret those features of life that we do observe as universal.

While the deeper questions about the ontological role of laws were largely left unresolved, a serious attempt was made to account for the specific universal features of life that are simplest and most primitive, for which the predictive power of biological laws should most resemble that in physics and chemistry. For these very old features, universal occurrence is more likely to indicate that few solutions to biological "function" were possible, and that this is why we have the forms we do. Understanding these structures is also likely to be critical as we try to piece together the origin of living from nonliving matter.

Such a focus on early core chemistry leaves many aspects of biological law unexplored, and even leaves us unable to say anything new about a host of regularities that the group examined, such as the beautiful weblike cell wall that encrusts all bacteria like a Fabergé egg or the ubiquitous scaling laws in ecology. However, with chemical universals as starting points, the group was able to embed biology in the larger geochemical world and also to look for the first place where uniquely biological forms of necessity differ from those in physics and chemistry.

A New View of Life's History

A lot has been learned about the earth's early geochemistry and the metabolic history of organisms since the early "chicken soup" models of the emergence of life. In the 1950s, Stanley Miller and Harold Urey showed that a broth of surprisingly complex molecules could be produced from the action of lightning in an atmosphere of ammonia, methane, and water, and this spawned a whole generation of models for the first emergence of proteins, DNA and RNA, and how these might have assembled into the machinery of cells (Miller and Orgel 1974).

The investigations were truly revolutionary, because they turned questions about the origin of life into laboratory science, and many of the experiments uncovered valuable pathways for synthesis of these important classes of molecules. At the same time, the enterprise fundamentally lacked structure, and no convincing overall origin stories ever came of it. It tried to account for the detailed combinations of molecules we see today but could give no deep reasons for why those molecules were important. Was containment in cells necessary to run the metabolism that traps energy to build

PHASE II: UNIFIERS

the cells? Were template molecules like RNA or proteins needed to select the reactions that would then build more RNA or protein? Each of these questions led to a version of the uniquely biological conundrum: "Which came first, the chicken or the egg?"

We are finding now that the four-billion-year history of life is divided chemically into two great periods, each about two billion years long. In the first, while sunlight was prevalent as it is today, living things appear not to have used it the way modern plants or blue-green algae do, and perhaps not at all. They may have drawn all of their power and material from energy-rich molecules bubbling up from volcanoes beneath the oceans. The molecules are simple and familiar—carbon dioxide, molecular hydrogen, carbonic acid (tonic water), hydrogen sulfide (rotten-egg gas), acetic acid (vinegar), or ammonia—but the realization that it is possible to live on those has only followed the discovery of families of modern deep-ocean bacteria that do just that. These remarkable organisms need nothing to eat besides such small molecules and inorganic mineral salts and can build all of their complex biomass literally "from the ground up."

Molecules bubbling up from magma are a limited resource, though, and it appears that photosynthesis emerged as a way to trap light to increase this resource pool, as certain purple bacteria do today. Only as a byproduct of storing energy from light in sugars did bacteria first produce oxygen, which could be used later to burn the same sugars to extract the stored energy. The large-scale adoption of this process converted Earth's atmosphere to the oxygen-rich form we know today, and introduced a whole new way of life, powered by eating sugars and metabolizing them with atmospheric oxygen. This was the second great period, in which life expanded to fill every niche on the earth's surface. We have traditionally viewed life inappropriately as if this were its only stage, simply because oxygen renders the older way of life impossible in the surface world where we live.

Chapter 12: Searching for the Laws of Life

Life Through the Looking Glass

When things happen in a particular order in history, it is often because the later stages build on the accomplishments of the earlier ones. The very fact that the early origin stories repeatedly run aground on chicken-and-egg paradoxes, where none of the steps seems possible before the others, suggests that we should take the two-stage history of life as an important clue. Can it be that the history of life is also a key to the emergence of the complexity of life?

The results of many different streams of work presented in the "Laws of Life" meetings suggest that this is indeed the case. For the last twenty years, Morowitz (1992) has been steadily rearranging the metabolic chart of all modern organisms, showing that the chemoautotrophic[1] reactions creating all the major classes of biomolecules originate somewhere on a single reaction cycle through ten compounds, known as the citric acid cycle, or Krebs cycle. This observation in itself is compelling, because it shows that the Krebs cycle is a kind of core of synthesis for all of biomass. However, that observation only goes part of the way toward simplifying our view of modern organisms, because for them the Krebs cycle is simply a way to digest sugars with oxygen, to produce energy. The energy digested is not used with the cycle compounds in any direct way to make biomass, and there is no obvious reason the chemicals in that particular cycle should be the starting compounds from which the rest of life is built.

Even more puzzling, the sugars digested by the Krebs cycle are now produced in plants by a separate complicated photosynthetic pathway, involving chlorophyll and many complex structures for managing energy and carbon flow. The molecules that perform photosynthesis perform no direct steps in their own replacement; that all comes from the Krebs cycle. No subset of this complex network

---

[1] *Autotrophs* are the self-starters of the food chain. The chemoautotrophs are organisms that require no intake of either organic matter or light in order to live. They self-sufficiently generate all of their biomass from inorganic small molecules.

PHASE II: UNIFIERS

of reactions can persist in isolation, because the whole network is required to supply any one of its compounds. At the molecular level, we again encounter chickens and their eggs.

The Krebs cycle, though, contains a telltale clue that modern organisms are not the place to look for its explanation. It is a cycle that takes two three-carbon sugar fragments, digests one of them to carbon dioxide to make energetic hydrogen ions, and returns the other through a so-called anaplerotic[2] reaction chain as a seed to begin the cycle again. Since the sugar is provided externally, this two-into-one cyclicity is not needed to supply materials. Further, from the perspective of SFI science, it is the ultimate paradox. The business of life is to build more life; why is the core engine of synthesis of all life a cycle that turns two copies of a complex molecule into only one? For many decades, a staple in abstract models of the emergence of complexity has been a self-catalyzed reaction that takes in one complex object and spits out two. The simplest such reaction, of course, is a cycle. The Krebs cycle has all the topology of such a so-called autocatalytic pathway (Fontana and Buss 1994)—only the reactions run around the cycle in the wrong direction.

The two-stage history of life resolves this deep puzzle, because in the earlier stage, the core biochemistry was a sort of mirror image of what we find in the later stage. We now know that the Krebs cycle is also present in the self-sufficient organisms of the first phase and is an engine of synthesis in them, just as in us. This is certainly true for the deep-ocean bacteria found today, and we suspect it has been a property of organisms back to the first cells. In these organisms, though, it runs in the right direction for autocatalysis. In other words, this reverse-direction Krebs cycle regenerates itself from nothing more than environmental small molecules, and then serves

---

[2] *Anaplerotic* is a term coined to refer specifically to the pathways that direct carbon in the modern citric acid cycle into compounds other than fully degraded carbon dioxide. These include the pathways that return it through oxaloacetate, an organic acid, to seed the next round of the cycle.

as a foundation from which all the rest of biomass is formed. Only after a complex life evolved to use and share sugars was an alternate pathway found to use sunlight for their formation. Then, the same Krebs cycle that had once built them was the most natural pathway to run in reverse, to break them down.

The centrality of the citric acid cycle in the metabolic chart suddenly makes sense, and in the autocatalytic direction, it no longer requires complex external pathways for the production of complex "food" molecules. Since the cycle itself is simple, involving only ten small compounds of carbon, hydrogen, and oxygen, it is also plausible as a primordial structure. This view is strengthened when one studies the internal chemistry of the Krebs cycle reactions, because it actually requires only three types of reactions involving C, H, O, and helper molecules like pyrophosphoric acid and hydrogen-sulfur molecules that may be available in some deep-ocean environments. The rest of the cycle, chemically speaking, comes for free. The important experiments that will be needed to see how the cycle relates to the origin of life involve how the reactions proceed without enzymes. Because modern organisms are highly optimized and use enzymes to fine-tune every internal reaction, there is no easy reconstruction of pre-enzymatic history from them.

Rewriting Origin Stories

These observations about the history of life and the structure of metabolism suggest a new family of scenarios for the origin of life, which is both structured and clearly lawlike, in comparison with the chicken soup scenarios. We envision that the earliest life was more like the self-sufficient deep-ocean bacteria than like anything else we know today. It formed around whatever chemical reaction cycles could convert the energy-rich but simple carbon- and hydrogen-containing molecules into structures that would seed their further consumption. The simplest such structure was the reversed

PHASE II: UNIFIERS

Krebs cycle, whose emergence and stability were driven by this metabolic capacity.

For any chemical mixture not tightly regulated by catalysis, the physics and chemistry of finite temperatures ensure that there is a cloud of surrounding reactions breaking down the chemicals toward lowest energy forms. When the starting chemicals are the intermediates of the citric acid cycle, these surrounding reactions contain the basic building blocks of sugars, fats, and amino acids that create proteins and nucleic acids that create DNA and RNA. They also contain the fundamental platelike molecules that are assembled to make chlorophyll, heme (which is wrapped in different proteins to make myoglobin and hemoglobin), and most of the metal-containing vitamins.

The story is completely reversed from the early scenarios of Miller and Urey. Rather than depend on relatively low concentrations of complex, atmospherically produced molecules, we expect that life originated in a steady, reliable environment that was relatively rich with simple but energetic molecules, as has been suggested by John Corliss (n.d.) and Günter Wächtershäuser (1988). The stability of their chemistry slowly led to a simple but stable non-background chemistry, in which carbon cycling in reverse through the citric acid compounds carried energy from the small-molecule "food" to equally simple small-molecule "waste," accumulating excesses of the building blocks of biomass as a byproduct. Stability of this rudimentary metabolism was the foundation for long trial and error, in which there was time for the discovery of those useful byproducts that could feed back to enhance the core metabolism, like the polar lipids (partly water soluble, partly oily molecules) that made cell membranes or the proteinlike molecules that are the simplest catalysts. These late-stage successes were not required to support the first metabolism or to provide the supply that allowed the experimentation to go on.

The thing that makes this origin story lawlike, however crudely, is its reliance on the ability to sample over and over again from a chemically "ordinary" environment. Rather than rely on chance reactions among rare molecules, it describes an emergent sequence in which each new level was available to be found, discarded, and found again, in samples from a stable level of structure directly beneath it. Metabolism selected from small-molecule chemistry. Biomass synthesized from a chemically stable metabolism. At each stage, the feature that emerged was the most stable, or most probable, that could be built on the foundation directly beneath it. The ability to identify structures as preferred, even in this probabilistic sense, embeds the lowest levels of biology in chemistry and physics.

What About the Genome?

One of the striking sociological features of biology today is the extraordinary importance placed on the sequencing and interpretation of DNA. The search for chemical regularity in the working group's discussions hearkens back to an older, even pre-Darwinian view of cause for the order of life. The older view says that living things have the shapes they have because, in some absolute sense, those shapes are good for something.

The early theories of visible characteristics[3] were often motivated by social, religious, or political idealisms, and gradually took on an aura of disrepute as scientific argument became more mechanistic, and (some) political ideals more egalitarian. During this transition, Darwin articulated the idea that inheritance with random variation determines what is possible, and competition then selects among the choices it is offered. Where the older arguments for "good shapes" seemed reasonable, Darwin's natural selection offered a way to converge on them, but the original notion of efficient design as a driving force toward good shape was lost in this transition.

---

3 Called *Phenetics*, from Greek *phainein*, to show.

PHASE II: UNIFIERS

In the century since Darwin, the first simple models of fitness with respect to an unchanging environment have given way to more subtle models, recognizing that species create each other's environments and so coevolve, and mathematical treatments have also made us more aware of how few of the possible forms and ecologies can ever be discovered at random. The idea of efficient design has thus been weakened even further—good solutions to problems can easily go undiscovered, and "Red Queen" dynamics of coevolution can cause all the species in an ecology to change in order to keep up with each other, while none of them actually "improves" in any obvious sense of its relation to the environment. The modern biological perspective is much more like one in which, to the genome, "everything is permitted," and the history of life is simply the history of accidents in the absurd races among genes in ecologies.

What does the historical record of the genome say about a metabolism-centric view of life, and about the role of design more generally? When only the genetic history of ribosomes (tiny bodies within cells that build proteins using information transcribed from DNA) had been reconstructed, there appeared to be a clear picture of the family tree of all life. Three major lineages lead to all of the modern organisms, of which two are types of bacteria. The third lineage contains everything else from yeasts to plants and animals. The cleanness of this description led biologists to expect that when a different genealogy from the DNA of nuclei was reconstructed, it would reinforce this ribosomal family tree and add detail to its earliest divisions.

What happened was rather different (Morowitz 2003). The DNA record muddies the early branches of the family tree by showing that the early single-celled organisms tried many different strategies for regulating their core biochemistry and exchanged the DNA that encoded these strategies rather freely across the early family boundaries. The three families still make sense, as identified by strategies for making structural walls and membranes of different types. Moreover, they seem to have all shared the core metabolism discussed above. These chemical features seem more stable, though, than the DNA that determined their regulatory machinery, as if chemistry determined the "right answers" to the cells' problems of metabolism and gross structure, and the DNA largely reified those right answers.

It appears as if the chemical "configuration" of the cells determined these earliest levels of structure more specifically than the genome did. If this is true, it suggests a change in the emphasis of biology, where absolute preferences for configuration interact with the mutation and selection of the genome to determine which forms of life can emerge and persist, and which cannot. We are not overturning Darwin's arguments about variation with selection, or returning to the Victorian notions of efficient design. However, we are learning to recognize that genes need not be rigid commitments, for good or ill, and that there may be aspects of life whose form is uniquely determined by the same sorts of thorough sampling that enable us to make specific predictions in physics or chemistry, an expectation that biology seems largely to have lost.

Physical Self-Organization and Biological Law

The group's investigation of primordial metabolism is very much in the spirit of studies of self-organization that have been traditional at SFI. Indeed, the emergence of an autocatalytic metabolism before there were enzymes, if it can be demonstrated, is as much a problem in pure physical chemistry as in biology. Yet clearly cellular life is more than pre-enzymatic metabolism, and biology obeys rules of order beyond those studied in physics. What light, if any, does our study of origin stories shed on these?

A focus on an emergent and self-sustaining core, from which living matter is constructed, alters our view of the many layers of

PHASE II: UNIFIERS

complexity that surround that core in all modern organisms. We see that enzymes for core reactions "pay their way" in improving the efficiency of the metabolic cycles that built them. Photosynthesis enabled primitive, volcanic metabolism to expand and fill the world by wrapping that metabolism in a chemical "spacesuit," which could generate food molecules from light. This freed organisms to leave the immediate neighborhoods of hydrothermal vents where they had evolved. Since the photosynthesizing molecules were themselves built from chemicals in the core metabolism, in augmenting it they provided for their own reconstruction.

At all levels of complexity in life, we see a hierarchical structure in which higher, regulatory structures sharpen or direct lower-level constructive processes. When they serve to enhance the processes by which they are created, either individually or cooperatively, they are favored by selection and become the stable innovations of evolution. Ultimately, every structure experiences some positive or negative bias from its impact on the energy and material extractable from core metabolism. This principle of construction may have no counterpart in physics, and yet may remain lawlike and predictive in the biological realm. It is like a feedback between components in a system, except that it operates between levels in a hierarchy. One could call this reciprocity between construction and regulation a "feed-down" relation.

It is tempting to see feed-down as an input to selection all around us, even at the levels of economy and society. In the economy, many activities lead to production, but those that generate capital enable us to change our means of production. For this catalytic effect, we actively work to protect those ways of life that generate and use capital. Similarly, many early states have their origins in piracy, when the pirates realized that they could extract more from local populations by living among them and instituting cooperative public-works projects. Theft became taxation, and the pirates became rulers, but only when their rule led to innovations such as cooperative irrigation, fisheries management, or reduction in internal conflict. We use such criteria today to distinguish legitimate states from other forms, though we still understand only poorly how to encourage legitimacy. An intriguing problem for the future is to see how much of anatomy, ecology, and sociology can be accounted for in terms of feed-down reciprocity (perhaps including some properties of Gould's phyla) and what it can predict about future innovations and change.

The authors wish to thank our co-organizer, Jennifer Dunne, for reminding us that the laws of life are hierarchical and must look upward to ecology as well as downward to physics and chemistry.

REFERENCES CITED

Corliss, J. B. n.d. "Creation of Life in Submarine Hot Springs." Privately published.

Fontana, W., and L. W. Buss. 1994. "'The Arrival of the Fittest': Toward a Theory of Biological Organization." *Bulletin of Mathematical Biology* 56: 1–64.

Gould, S. J. 1989. *Wonderful Life*. New York: W. W. Norton.

Miller, S. L., and L. E. Orgel. 1974. *The Origins of Life on the Earth*. Englewood Cliffs, NJ: Prentice-Hall.

Morowitz, H. J. 2003. "Phenetics, a Born-Again Science." *Complexity* 8(1).

———. 1992. *Beginnings of Cellular Life*. New Haven: Yale University Press. (See also the feature on Morowitz in the Santa Fe Institute *Bulletin*, vol. 15, no.1, Spring 2000, 2–5.)

Wächtershäuser, G. 1988. "Before Enzymes and Templates: A Theory of Surface Metabolism." *Microbiological Reviews* 52(4): 452–84.

METAPHORS: LADDERS OF INNOVATION

David Gray and Michele Macready
Boston Consulting Group
SFI Bulletin, Winter 2004

Though often dismissed as mere rhetorical window dressing, metaphors play an important role in innovative thinking. In particular, the *cognitive* use of metaphor can reveal potentially fruitful connections and novel ways of seeing that lead to new insight. There are many modes of metaphorical thinking, and an analysis of its operation in science, as in other domains, requires attention to the intention of the metaphor, its essential structures, and the different types of impact it can produce.

A Necessary Ladder

A two-day workshop organized by the Santa Fe Institute and the Strategy Institute of the Boston Consulting Group (BCG) last April brought together practitioners and academics from a number of fields with a common interest in the topic of metaphor. SFI participants Walter Fontana, José Lobo, and Jim Rutt gave presentations on the use of metaphor in their respective areas of chemistry, economics, and business. Paul Humphreys and Nicholas de Monchaux of the University of Virginia presented, respectively, a philosophical account of metaphor and its use in shaping visions in architecture. Tiha von Ghyczy of the University of Virginia's business school and the BCG Strategy Institute, together with Michele Macready and David Gray, also of BCG, reported on the Strategy Institute's effort to build an online "gallery" of multidisciplinary metaphors to inspire business thinkers and reflected on the potential of employing large sets of metaphors as aids to creative thinking. The meeting

PHASE II: UNIFIERS

explored the use of the cognitive metaphor as an important element in innovation in all these disciplines. The road to novel theoretical work consistently winds through a forest of metaphors.

Complexity science is premised on the assumption that seemingly disparate phenomena, both natural and social, evolved and constructed, can be understood using a common conceptual framework. The signature concepts used to talk about complex systems—emergence, adaptation, networks, evolvability, phase transitions, self-organized criticality, fitness landscapes, robustness, learning, edge of chaos, even the very notion of complexity itself—remain more metaphorical and suggestive than definitional and precise. And how else could physicists, biologists, chemists, economists, anthropologists, ecologists, computer scientists, and historians engage in meaningful scientific dialogue without the ferocious exchange of metaphors?

> A useful metaphor is an invitation to hard work that can be indispensable to innovation.

Reliance on metaphors is by no means unique to complexity science, of course, but is instead prevalent in every field of scientific inquiry, especially in its early stages. Nor is the importance of metaphors confined to rarefied reasoning: the use of metaphors shapes our basic perception and understanding of the world. And yet scientists often distrust metaphors. Metaphors are not models and are thus not susceptible to the sort of direct application and rigorous testing that are the gold standard of scientific verification. As such, metaphors are sometimes viewed as incomplete—or worse, shoddy thinking. While acknowledging their appeal, many regard metaphors merely as ladders that, to paraphrase Ludwig Wittgenstein

Chapter 13: Metaphors: Ladders of Innovation

(no slouch himself when it came to the use of metaphors), once used to climb to a conceptually novel place must then be discarded.

At SFI, concerns for the proper role of metaphors and a respect for the difficulties in transitioning from metaphors to models have been present from the beginning and continue to animate discussion, from the 1992 "Integrative Themes Workshop"[1] to a recent workshop on the "Robustness of Coupled Natural and Human Systems." Plenty of Wittgensteinian ladders continue to be erected and kicked away in complexity science. At the Strategy Institute, the cognitive use of metaphors in developing innovative strategies has been at the center of recent work.[2] The insights gained have already started to make an impact on practical work for clients.

Yet, the prevailing view systematically underappreciates the critical operation of metaphor in cognition—whether in science, the arts, or business. A metaphor is not merely a flawed and fuzzy model, nor is it a final answer. A useful metaphor is an *invitation to hard work* that can be indispensable to innovation. Metaphors and models are not locked in a battle for relevance but can be seen as successive ladders, stacked one upon the other, which continue to underpin good thinking. W. Brian Arthur acknowledged this state of affairs at an SFI conference a few years ago when he said: "I have a very strong belief that science and thinking progresses not so much by theorems but by metaphors. Metaphors are what we absorb, that go in deep, that we digest, perhaps also consciously forget. But two years later you start to write about evolution in the economy and (suddenly you find yourself) deeply informed about how it takes place."

One goal of the April workshop was to discern some of the essential aspects of metaphors that make such unlikely, playful

---

[1] See George Cowan, David Pines, and David Meltzer, eds., *Complexity: Metaphors, Models and Reality* (New York: Addison-Wesley Publishing Company, 1994).

[2] See Tihamer von Ghyczy, "The Fruitful Flaws of Strategy Metaphors," *Harvard Business Review* (September 2003): 86–94, for a discussion of cognitive metaphors in business innovation.

PHASE II: UNIFIERS

connections so highly productive. What constitutes the "appropriateness" of a metaphor, and where do good metaphors come from? More fundamentally, are there ways to improve our prowess as metaphorical thinkers, and can the novel topologies created by the mixing of metaphors, such as occurs regularly in crossdisciplinary work at both institutes, increase their power? This paper is an attempt to address some of these thorny questions and draws heavily upon conference presentations for insights and examples.

Metaphorical Reasoning

Metaphors appear almost everywhere in our conscious experience. While the classical use of the term applies primarily to a literary device, metaphors can also be visual or even auditory or olfactory (say, comparing something to the aroma of baking bread or apple pie). The organization of the computer "desktop," with its folders and file cabinets, is built on a metaphor. And what are we to make of so-called "metaphors of use" that let us see that, for certain purposes, a dime is a screwdriver? What is it that makes all these things potential metaphors?

> A liaison that does not involve some transgression of boundaries is no metaphor.

"A metaphor," writes philosopher Nelson Goodman, "is an affair between a predicate with a past and an object that yields while protesting."[3] This rather louche, metaphorical definition highlights an essential feature of the metaphor: an intrusion from one domain into another. The metaphor borrows language, symbols, logic, and

---

[3] Nelson Goodman, *Languages of Art: An Approach to a Theory of Symbols* (Indianapolis, IN: Hackett Publishing Company Inc., 1976), 69.

Chapter 13: Metaphors: Ladders of Innovation

associations from one field and imposes them upon another to which they do not properly belong. Thus, the notion of a "war on poverty" suggests a transfer of structures and associations from the military domain to the social. The fit may be uncomfortable and endured only under protest, as Goodman notes. It is the incommensurability of the metaphor that is often its salient characteristic: if taken literally, all metaphors are patently false if not absurd. Melville's declaration that "Christ was a chronometer"[4] is a stark example. A liaison that does not involve some transgression of boundaries is no metaphor.

Thus, the fit of the metaphor will always be inexact, and it is around these jagged edges that much of its innovative potential lies. A metaphorical intrusion smoothed by time and long wear is apt to become a dead metaphor or cliché: "Achilles heel" no longer jangles with associations of its original source but is used quite unreflectively to denote a fatal weakness. For most in the business world, the exhortation to "think outside the box" evokes no connection to the brainteaser that spawned the phrase. These metaphors have died into literalness and thus lost their power to catalyze thinking.

We should note that language is thick with the corpses of dead and dying metaphors. Scratching the etymological surface of most words reveals their metaphorical roots: the "corporation" derives from the *corpus*—a living body—and "strategy" from *strategos*—a military general. We are quite justified in using words literally without constantly acknowledging the underlying metaphors, but unearthing these foundations can sometimes be revealing. Unquestioned, implicit metaphors continue to exert a strong effect on the structure of language and thus on the structure of thought itself. For instance, bringing to light the mechanical metaphor implicitly embedded in a lot of business thinking (which continues to spawn new submetaphors, like "alignment," "toolkits," and

---

[4] Herman Melville, *Pierre: Or, the Ambiguities* (New York: Harper & Brothers 1852).

PHASE II: UNIFIERS

 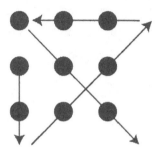

EXERCISE 1: The player must connect all 9 dots with four straight lines without lifting the pencil from the page. The solution requires thinking outside the (nonexistent) box created by the square arrangement of the dots.

"reengineering") can cause us to reconsider whether we are operating with the right picture in mind. Attention to metaphor allows us to engage in a useful archaeology of clichés.

Anatomy of a Metaphor

Linguists use the terms *source* and *target* to designate the linked domains of the metaphor. The target is the main topic of discourse (e.g., the development of scientific ideas)—the thing we wish to understand—and the source is the interpretive device that sheds light on the target (e.g., political revolutions). Typically, we would expect the source domain to be the more familiar to us, the one closer to understanding or intuition, which therefore allows it to elucidate the more obscure target.

Many metaphors, however, draw upon source domains of considerable complexity: for example, laminar flow as a metaphor for business supply chains. Among business practitioners, the invocation of laminar flow is likely to produce a lot of blank faces, while the supply chain (itself a metaphor!) will be quite familiar. In this case, the effectiveness of the metaphor is not immediate but requires a great deal of education to make it work.

The example highlights an important feature of metaphor—its power to *defamiliarize* the familiar. We think we know something about supply chains: the interlinked system of companies, individuals, and goods that provides inputs to manufactured products. The effectiveness of the metaphor borrowed from physics lies in its power to unhinge this knowledge—is it a chain? or is it more like a smooth flow of liquid? or is it a web?—in a way that allows new thinking to penetrate. We need not discard the existing picture, but the effective metaphor causes us to add new dimensions to the conceptual space. We may, therefore, wish to replace the notions of "familiar" and "unfamiliar," substituting "known" and "unknown." In some sense, the cognitive flow of the metaphor will always be from a domain of knowledge to

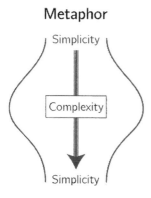
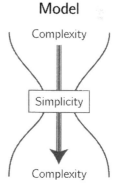

FIGURE 1

"I would not give a fig for the simplicity this side of complexity, but I would give my life for the simplicity on the other side of complexity." —Oliver Wendell Holmes, Jr.

"Frustra fit per plura quod potest fieri per pauciora." —Occam's Razor

"Everything should be made as simple as possible, but not simpler."—Albert Einstein

PHASE II: UNIFIERS

one of nescience, but this does not necessarily correspond to the intuitive familiarity of these realms.

We should note that the transfer (the word *metaphor* itself comes from the Greek roots meaning "to carry across," i.e., transfer) can work in both directions. The linking of neurophysiology and computation is a commonly cited example of such a "boomerang" metaphor. While the initial borrowing of language and concepts flowed from brains to computers, the favor has been returned in the form of computer theory as a source of metaphors for neurological processing and, more generally, for information processing in biological processes. It may be more proper to speak of "ricochet" metaphors—once fired off, the trajectory and related combustions touched off by the cognitive metaphor may be difficult to predict! Darwin was clearly influenced by the works of political economists like Malthus and Smith in developing his principle of natural selection. A century and a half later, we see the emergence of an army of researchers eagerly applying biological insights to the workings of markets. Likewise, the authors of a new book[5] relating the strategic insights of military theorist Carl von Clausewitz to business note that they are merely returning the metaphor: Clausewitz himself proposed that war could best be compared with commerce, since both are social conflicts of human interests and activities.

Metaphor and Analogy

Analogy is closely linked with metaphor and, indeed, we may think of these two notions as interchangeable: both point out likenesses in particulars between things that are otherwise dissimilar. In the way we propose to use the term, *analogy* is a component of metaphor

---

5 See Tiha von Ghyczy, Bolko Oetinger, and Christopher Bassford, eds., *Clausewitz on Strategy: Inspiration and Insight from a Master Strategist: A Publication of the Strategy Institute of the Boston Consulting Group* (New York: John Wiley & Sons, Inc., 2001).

that refers to the correspondences between domains—the structures or associations that form the core of the link between source and target. Without some degree of analogical mapping, the metaphor will be stillborn. A metaphor, however, goes beyond analogy by including all the ill-fitting facets of the linked entities—the fractures and fault lines—in the picture. The metaphor comes to life where analogy leaves off.

At this point, we raise again the vexed question of metaphors and models. In some cases we want to make a sharp distinction between these two things, while in others they seem to live in harmony. The question arises with particular force in the sciences where metaphors (e.g., plum puddings or solar systems as images of the atom) seem to shade into models that shape experimental design. The metaphorical origins of scientific models have been long noted: Perhaps every science must start with metaphor and end with algebra, and perhaps without the metaphor there would have been no algebra.

Yet models have certain recognizable properties distinct from metaphors, most notably a degree of formalization that the metaphor lacks. We want models for specific purposes, and we demand of them a certain rigidity that preserves the essential relationships between the model and the modeled. We might ask whether the mechanism of the model necessarily involves an appeal to the formalization of mathematics. A ship model, for example, provides a formal mapping to the actual bark by means of a mathematical correspondence, say 1 centimeter = 1 meter. A financial model similarly purports to capture the essential activities of a firm and their relations to one another using the formal structures of mathematics.

Figure 1 illustrates the differing intentions of model and metaphor. The widening body on the left represents the "complexification" accomplished by the metaphor. For example, "Juliet is the

PHASE II: UNIFIERS

sun" brings in a host of potential structures and associations (e.g., of light, warmth, rising and setting, perhaps also eclipse?) that vastly complicate our picture of an otherwise unremarkable teenager from Verona and leaves that object forever altered in our understanding. The pinched figure on the right represents the opposed mechanism of the model that seeks to make tractable a potentially vast body of inputs and data through necessary simplification. This simplification is far from the final word, however, as analysis of results gives birth to new types of complexity.

This picture may seem to imply a fundamental opposition between metaphor and model. But this need not be so. One attempt to cut this Gordian knot is simply to identify models with analogy, which would lead us to say that a model is a special kind of metaphor—one that has been pared down to the hard core of direct correspondence we can map between the two domains. This leaves the door open to further revision of the model through a metaphorical process but also lets us make formal demands of the model that may lead to its "falsification" or rejection if it fails to deliver on its promised correspondence to reality. Figure 2 is an attempted illustration of this relationship that stresses the analogical properties

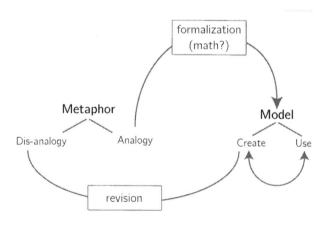

FIGURE 2: Metaphor and Model

of the model while allowing for an ongoing reciprocity between model and metaphor.

Dialectical Taxonomy

The metaphor appears in so many guises and incarnations that a formal representation of its workings would seem very nearly futile. Instead, our analysis leaves intact all the inherent tensions. This description incorporates a sort of dialectical pairing of aspects of metaphor without attempting an ultimate synthesis. The progression is described in three stages: the intention, the structure, and the impact of the metaphor.

Intention of the metaphor

In large part, intention makes the metaphor. There is no reason a priori why two such unrelated domains as ant foraging behavior and airline baggage handling should ever meet. The metaphor arises in the deliberate pairing of these two things. Although some things strike us as *found* metaphors, the fact and the nature of the linkage derives more from the intention of the metaphor maker than from inherent structures. Intent must be appropriate to the context in which the metaphor is used and largely determines its success. Thus, the particular application of metaphor in the business sphere in which the Strategy Institute operates may differ from its use in the SFI context of scientific research.

Much of the discussion of literary metaphors, going back to Aristotle, deals with them as *rhetorical* devices whose purpose is to convince the reader through an especially apt linkage. The rhetorical metaphor relies on economy of expression and aesthetic impact. The source domain of the rhetorical metaphor should be familiar and intuitive, and the analogy between the linked entities immediately compelling. We distinguish from this the *cognitive* use of metaphor that forces a departure from the familiar. The intent of the cognitive metaphor is

PHASE II: UNIFIERS

novelty, and we encounter it primarily as an opening to further inquiry. The linkages may be far from intuitive—for example, a proposed rapport between the intangible structures of proof theory in logic and chemistry.[6] But the intention is to force a cognitive reevaluation that lets us see in new ways and ask different questions.

Thus, two alternative intents of the metaphor are revealed: *invitation* and *persuasion*. We are invited by the cognitive metaphor to delve into the intricacies of the source domain, test the strength of the bridge to the target, and hunt for important fractures in the analogy. It is an invitation to become a co-creator of the metaphor. The rhetorical metaphor is more concerned with persuading us to a certain view.

This points to a further pairing of intentions related to the use of metaphor—*synchronizing* and *disarming*. Metaphors often become shorthand for certain ideas and are thus useful in aligning understanding or expectations. A manager who uses military terminology to describe business situations takes advantage of the synchronizing power of the metaphor, so that the difference between a "flank maneuver" and a "frontal assault" on a competitor is quickly understood. Alternatively, metaphor can disarm expectations—we may choose the metaphor of laminar flow or neo-Darwinian evolution precisely because it is unfamiliar. With the metaphor comes an unaccustomed vocabulary for describing phenomena in the target domain that forces a reexamination of what we know.

A final dimension of intention involves the use of the metaphor for *creating* versus *distilling* knowledge. The purpose may be to produce a new lens that allows an alternative view of the area we seek to understand. Here, the disjunctions and fault lines of the metaphor can be especially productive and drive us to search out or

---

[6] See Walter Fontana and Leo Buss, "The Barrier of Objects: From Dynamical Systems to Bounded Organizations," in *Boundaries and Barriers: On the Limits to Scientific Knowledge*, eds. John Casti and Anders Karlqvist (Redding, MA: Addison–Wesley, 1996), 56–116.

create structures that appear to be missing in the mapping between source and target domains. The distilling function of metaphor, on the other hand, is less about creating new knowledge than encapsulating wisdom—often gained through long experience—into a form that can be communicated to others.

These dichotomies reflect two categories of metaphorical intentions: learning and communicating. While the distinction is not absolute—we may learn much in seeking to communicate insights through metaphor—there is a natural split in the foregoing pairings. The learning function of metaphor emphasizes extended engagement with the source domain as a way of shaking up received thinking. It is more likely to focus on the fractured edges of the analogy, seeking novelty in the interstices where the fit is most uncomfortable. Rhetorical, persuasive, synchronizing, and distilling uses of metaphor are more geared toward capturing and communicating subtle insight to others.

Structure of the metaphor

The building blocks of metaphor—source, target, mapping, analogy, fracture—have been discussed in some detail above. But a number of tensions surrounding the structure of metaphor arise in its application. One such tension pertains to what might be called the "level" of the metaphor: Does it act as a *governing* paradigm or in an illustrative, *subsidiary* role? A grand, governing metaphor offers a holistic interpretation of the target domain. Some governing metaphors may be unacknowledged—though no less extensive or influential for that reason. Implicit images of firms as machines or as organisms are pervasive in business discourse[7] and shape our understanding and expectations of action, authority, and change. But not all metaphors make such claims to completeness. When

---

7 See Gareth Morgan, *Images of Organization*, 2nd ed. (Thousand Oaks, CA: Sage Publications, 1997).

PHASE II: UNIFIERS

Wittgenstein uses a toolbox to illustrate the diversity of language, or Adam Smith speaks of the intercession of an invisible hand in the market, our enlightenment does not depend upon acceptance of a larger schema.

Indeed, we often find that effective metaphorical thinking involves not just one *grand schema* but a *mixing of metaphors*.

> We use the metaphor as a kind of "wrapper" to give us a mental grip on a slippery substance.

While such promiscuity is deemed poor style in the literary metaphor, its cognitive use is enriched by a proliferation of viewpoints. This raises some fascinating questions about the topology of large sets of metaphors: the strength, valence, mutability of connections among the source domains. The metaphorical space suddenly expands geometrically. What is lost, perhaps, is the coherence that a single schema, rooted in a particular domain, makes possible. Again, the intent of the metaphor—as a discovery device versus a communication tool—may dictate the effective structure.

A final reflection on the structure of metaphor deals with the trade-off between *depth* and *shallowness*. The domain acting as the source for the metaphor is rarely taken in all its complexity but is, at best, a snapshot—a frozen picture that provides the basis for the metaphorical transfer. Thus, the structures of biological evolution have been taken seriously as a metaphor in economics at least since the 1950s, but the picture of evolution that economists work with is typically limited.[8] Our understanding of the source can and should be continually revisited and revised. But to be effective for the metaphor, it must have an *appropriately moderate* number of dimensions. Too much complexity renders the metaphor unmanageable. For the pedantic astrophysicist intent on solar flares and burning helium, the comparison of Juliet to the sun will be unrevealing. There is a sense in which all metaphors are shallow and must remain so. We can seek expertise in the target domain, but, in the metaphor, we approach the source domain as amateurs.

But as *curious* amateurs! Unlike the model, which demands a degree of closure and completeness, the process of metaphor has no defined terminus. The original evolutionary metaphor in economics may be updated to include empirical manifestations of epistasis in fitness landscapes or insights from the sequencing of the genome. These revisions may provide valuable extensions of the metaphor—although this is by no means guaranteed. The depth of the metaphor lies in its open-endedness.

Impact of the metaphor

The outcome of a metaphor will partially depend on its original intent. In some cases, the metaphor may prove immediately effective, while in others considerable effort may be required for it to bear fruit. Its impact may come either in "working" the metaphor or in *using* it. To use a metaphor means to apply the insights, language, equivalences, and other associations of the source to shed light on the target. We do this all the time. Generations of physics students have used the familiar notion of water moving through pipes as a way of conceptualizing the much more intangible flow of electrical current. The parallels break down at some point, but it is

---

[8] Armen Alchian, "Uncertainty, Evolution, and Economic Theory," *The Journal of Political Economy* 58, no. 4 (June 1950): 211–21, is widely regarded as an early influential article in this vein.

PHASE II: UNIFIERS

a useful early device for learning. We use the metaphor as a kind of "wrapper" to give us a mental grip on a slippery substance.

The effect of working the metaphor is rather different. The impact comes more in its creation than in its eventual application. Working the metaphor means plunging into the intricacies of source and target domain and building the bridge between them span by span. The metaphor that emerges may not be intuitive or easily applicable (e.g., *NKC* landscapes as images of economic ecologies), but the process of generating it may trigger unanticipated insights. The benefit arises from the different perspective one adopts in plumbing the intricacies of the metaphor.

This also raises the question of just who is using or working the metaphor: that is, the metaphor's community. Is the metaphor the property of an *individual* or of a *group*? Take, for example, a metaphor that publicly shaped US foreign policy for decades: the "domino effect." This theory expressed the fear following World War II that, if one country were to fall to Communism, its neighbors would fall with it (like a row of dominoes). Whatever the merits or limitations of this mechanistic metaphor, there is no doubt that it had many adherents who used it in forming and communicating ideas. Creation and refinement of a metaphor is often the work of a group. The emerging model of the atom in the first half of the twentieth century benefited from the metaphorical contributions of multiple minds: Thomson's plum pudding, Rutherford's solar system, Bohr's water droplet, and so on, that invited a whole generation of physicists to continue this theoretical exploration.

On the other hand, there are private metaphors that lead to insight for one particular mind. An example of this type of heroic metaphor might be Albert Einstein's thought experiment in which he imagined how the world would look to him riding a beam of light (as if it were a train or a horse). The change of perspective that the metaphor allowed made possible the later development of his theories of relativity without demanding, however, that others adopt the light-riding metaphor themselves.

While most of the discussion has been about the logical transfer of structures between target and source domains, an analysis of metaphors that did not take into account the emotional and intellectual associations that attach to them would be incomplete. Metaphors do not come without baggage, and their impact may have as much to do with these ancillary factors as with their formal content. Indeed, it has been proposed that the cascade of associations, both positive and negative, triggered by the metaphor is its content. These associations are not only inescapable but integral to the impact of the metaphor. For example, the effectiveness of military metaphors in business may have primarily to do with the penumbra of associations—camaraderie, loyalty, sacrifice, determination, and so on—that surround warriors. There is thus a dualism between the *logical content* and the *cascade of associations* of the metaphor in assessing its impact.

In highlighting the various strands and tensions of the metaphor, we have raised more questions than we have resolved. At the least, we hope to have made clear how pervasive metaphor is and how multifarious its use. In particular, we wish to recognize the essential role of *cognitive metaphors* in creative thinking. As with physical ladders, metaphors must be used with care, planted firmly, and adjusted to the task at hand. And whether we then quietly put them aside or continue to build on these edifices, we will always need to resort to ladders for climbing to new conceptual heights.

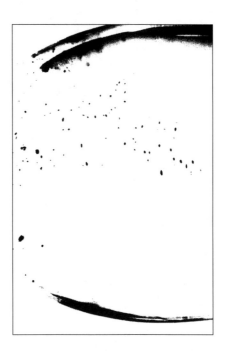

THE NUMBERS OF OUR NATURE: IS THERE A MATH OF STYLE?

Daniel Rockmore, Dartmouth University
SFI Bulletin, Winter 2007

In May 2005, Alex Matter, a filmmaker and son of photographer Herman Matter and painter Mercedes Matter, announced the discovery of thirty-two early Jackson Pollock drip paintings. The full story sounded a lot like a New York version of Michael Frayn's novel *Headlong*. Matter's parents were friends of Pollock, close enough to the artist that photographs exist of them lounging about with Pollock. At least one biography of Pollock mentions the purchase of some small Pollock works by the Matters, and it is reported that Herman Matter's signature is on the back of these works, labeling them as Pollock originals. It seems that the paintings had been languishing in a metal storage bin out on Long Island until they were discovered by Mr. Matter, not far from Pollock's former home in the Hamptons and the scene of his fatal car accident. The press releases were accompanied by pictures of a grinning Alex Matter holding one of these paintings, which, at a distance and at the resolution of a newspaper photograph, looked suitably spattered. It's a good story and certainly a plausible one.

But, in fact, hardly a year goes by in which someone doesn't claim to have found a long-lost Pollock—that's what Richard Taylor, a consultant for the Pollock–Krasner Foundation, tells me. Most of these claims are disposed of easily, their faults ranging from the obvious (misspelled signatures—"Jackson Pollack") to the subtle (materials that were only available after Pollock's death). Some fakes are veterans of the auction circuit and reappear every few years with a new story ("Some hitchhiker gave this to a truck driver, who brought it into my gallery..."), while other schemes are

PHASE II: UNIFIERS

almost as ingenious as the artist himself, even going to the extremes of inventing a provenance by inserting appropriately faded pictures of the "discovered" work into obscure hard-to-find gallery catalogs.

So, how to tell real from fake? In spite of the fact that there is only one right answer, generally, evaluating claims of authenticity in the world of modern art is, well, more art than science. Slam dunks are the situations in which the work is accompanied by an airtight provenance, consisting of documentation of every single person or institution that has ever owned the work (the back of a painting can look like a well-traveled steamer trunk, covered with labels marking its appearance in exhibitions). Such works are also accompanied by records of any conservation incurred and a list of events detailing the touch-up here or the cleaning there. Failing this sort of complete biography, the truth begins to get a little murkier. Sometimes science can help. X-rays can reveal underpaintings or preliminary sketches. Analysis of the materials provides clues, including information as to the age of the work. But even this hard evidence can go only so far. Several years ago the photography market was flooded with "newly discovered" Man Ray prints, their authenticity supported by the fact that the photographic paper was exactly of the kind that Man Ray was known to have used sixty years ago. It was only later discovered that the forger had managed to find an old cache of this paper, obtain access to Man Ray's negatives, and use this to make a new set of prints.

So, enter the human actor. Connoisseurs steeped in the work of the artist in question apply the "sniff test" and either the light bulb goes on or it doesn't. And the fact is that different connoisseurs of the same artist can have different senses of smell. Ultimately, in the case of Alex Matter's bundle of proposed Pollocks, the experts are taking a close look at each painting and answering the question, "Is this work characteristic of the style of Jackson Pollock at this stage in his career?" This is the question that Richard Taylor asked when the Pollock–Krasner Foundation called him in to evaluate the Matter cache—but the way in which he differs from his connoisseur colleagues is that he answered the question with mathematics. Taylor seems to have found a way to extract a numerical signature that identifies a work as an original Pollock. More precisely, Taylor has determined that a statistical analysis of the numbers that comprise the digital representation of the work can reliably separate authentic Pollocks from fakes. In so doing, Taylor's work is paving the way for a new world of visual "stylometry," that is, a growing discipline that is making math out of that seemingly impossible to define thing that is *style*. It is a subject that actually goes back to the mid-1800s but has recently exploded in our world in which all actions and creations are fodder for the computer and for those with broad vision. The tools of trade find their way into all aspects of our life, helping to distill not only styles of artistic creation but styles of all sorts of actions.

Measuring Style

Mathematical analysis or statistical analysis requires that something be measured, so it's not surprising that as applied to the arts it is able to first find a foothold in literature. Books, essays, any sort of written output, present a numerical collection, containing all kinds of primary data. Letters, words, sentences, and so on can be counted and basic statistics can be gathered: How often is a given word used? What is the distribution of word length? What is the distribution of sentence length? etc. Do this and you are guaranteed to get numbers. What is not guaranteed, and hence is surprising, is that out of these numbers, patterns emerge, both on the scale of society as well as the individual. In the former, we see the patterns that seem to be intrinsic to any form of communication, and in the latter, we seem to be able to distill aspects of the idiosyncratic patterns of usage that form the basis of a person's writing style.

PHASE II: UNIFIERS

The idea that mathematics might be useful for determining authorship is usually attributed to the nineteenth-century British mathematician Augustus de Morgan. De Morgan was in many respects ahead of his time, especially as regards what appears to have been a broad-based investigation of the power and possibility of the formalization of thought. An interest in the formal or quantifiable aspects of creative work easily fits into this program.

As recorded in his wife's memoirs, de Morgan wrote a letter to a friend in 1854 that states, "It has always run in my head that a little expenditure of money would settle questions of authorship" by determining whether the writings of "the latter do not deal in longer words than the writings of the former." It must have been the case that de Morgan never had the extra pocket change to pursue this thought, as it was several years later, in 1886, that the first published account of a mathematical analysis of literature appears, written by American physicist Thomas Mendenhall. The subject is Shakespeare, a favored focus of the question of attribution. Mendenhall tries to distinguish between Francis Bacon and William Shakespeare according to their relative use of four-letter words. It's not a success, but nevertheless, a discipline is born. It finds a name when ten years later, Wincenty Lutoslawski looks at five hundred numerical attributes in each of Plato's dialogues in order to reconstruct the order in which they were produced, working with the basic philosophy that works that are close mathematically should be close temporally. He called his methodology *stylometry*.

Modern trends in stylometry pull from the full bag of tricks of artificial intelligence and advanced statistical analysis. Some approaches focus on aspects of predictability—using the empirical likelihood that one word is followed by another. These ideas were first proposed in the early 1900s by the Russian mathematician A.A. Markov, who used them to construct a very simple model for the cadences found in Pushkin's poem *Eugene Onegin*. Today, so-called

Chapter 14: The Numbers of Our Nature

Markov chains are among the most commonly used tools in the mathematical modeler's work belt and can be found identifying patterns in all sorts of places, ranging from genetic to financial data.

Word frequencies used in one way or another remain the heart and soul of literary stylometry. Some of the most successful techniques focus on usage statistics of *function words*. As opposed to *content words*, these include pronouns, conjunctions, and prepositions, which generally carry very little noncontextual meaning and serve instead as grammatical connective glue. Literary style, it appears, resides in the degree to which we choose *that* rather than *which* or *however* as opposed to *nevertheless*. The starting material for many a stylometric analysis begins by first isolating and recording the frequencies of the favorite function words among works of known authorship and then, in one way or another, considering the degree to which the frequency pattern in a contested work is statistically similar to the patterns in the secure works. Among other examples, function word usage has been used to distinguish between the writings of Alexander Hamilton and James Madison as well as to pinpoint authorship in the *Wizard of Oz* series.

One of the most striking results in the field is the discovery that there are certain patterns of usage that seem to be simply intrinsic to the act of communication. In 1949, using a corpus of a range of works by a number of different authors, Harvard linguist George Zipf discovered a remarkable empirical fact, known today as Zipf's law. It states that the result of multiplying a given word frequency by its rank (the most frequent word has rank one, the second most frequent is rank two, and so on) was approximately the same over all the words in the corpus. Zipf published his analysis in an amazing book titled *Human Behavior and the Principle of Least Effort*, which derives its name from the basic argument that predicts the Zipfian discovery: Imagine any author as a chance-driven machine in which at each step a coin is tossed—if it lands heads up, then a previously

PHASE II: UNIFIERS

used word is chosen at random, while if it lands tails up, then a new word is written down. This is a "rich get richer" sort of model in which the more a word is used in the past, the more likely it is that it will attract more use in the future. In fact, Zipf finds similar relations (called *power laws*—a familiar distribution in the complex systems world) among all sorts of ranked lists, ranging from sizes of towns (here you would consider the product of the population of the town with its rank) to income distributions (where it also goes by the name of Pareto's law).

Words can be counted—that's the main reason that stylometry came first to literature. But other art forms have natural numbers, too. Given the success of literary stylometry and the empirical ubiquity of Zipfian behavior, it's something of a surprise that it was not until just a few years ago that a broad analysis of that other great symbolic language that is musical composition was undertaken. Charleston College computer scientist Bill Manaris led a small group of researchers that counted note usage over a range of composers and works, and discovered a basic power law structure. Using that as the foundation of his analysis, he and his colleagues were able to derive a collection of statistical features from musical scores that successfully allow an automatic classification of musical works from jazz, classical, and rock 'n' roll.

Actions Speak as Loudly as Words

Even with movement, it is possible to distill style from the numbers. In its measurement of angular displacements of joints and relative displacements of limbs, *kinematic analysis* reduces human movement to streams of numbers. Plug-and-play animation software is evidence of the fact that there is a basic mathematical formulation to the way in which we move—there is an average walk, run, jump, and so on. Stylometry comes into the picture as a means to give these virtual folks the style of movement of particular real people, accomplished by motion-capture systems (mocap) that can track and record the movements of a collection of sensors or transmitters worn about the body. The acquired data can provide a numerical record of the particular way that a person walks or sips tea—and then the animator can take that away, using the manner in which that person differs from the average to give a personal touch to an animated avatar. Motion capture is a modern updating of the way in which the early Disney animators worked—often tracing over stills gleaned from footage of the movements of professional dancers and clowns in order to acquire a feel for a basic movement style. Mocap is now a standard Hollywood technique, responsible for personified performances such as Tom Hanks's animated turn as the conductor in *Polar Express*. This Frankensteinian "retargeting" of the motions of the living to the lifeless has even been attempted from cartoon to cartoon. Through an adaptation of the mathematics of mocap, Fred Flintstone can take dance lessons from Mickey Mouse.

The success of motion capture for animation is evidence that actions do indeed speak at least as loudly as words. This is an idea that is driving a new generation of marketing and advertising. Your shopping has a style that is encoded in the trace of clicks and eyeball dwells that you leave on the web, your checkout list at the grocery store, and your monthly credit card bill. The electronic hectorings "If you liked this, then you'll like that!" are mathematical statements based on a geometry of sales space—an item that you buy is encoded as a list of numerical attributes, and like the two-number lists that make up the x, y coordinates studied in a high school geometry class, these lists of item coordinates give geometric meaning to your buying habits. If marketers find you shopping in one region of their abstract product space, they've a pretty good idea that you may like some items nearby that their specially designed math-marketing goggles allow them to see. This approach can even be applied to shopping for a mate—that's at least what some of the online dating

PHASE II: UNIFIERS

services rely on; your personal style has a shape, in the most mathematical of terms.

Painting by Numbers

That the visual arts have been among the last to embrace any form of stylometry is perhaps more a matter of tradition than anything else. As a discipline, connoisseurship has scientific origins, usually attributed to Giovanni Morelli, a nineteenth-century Swiss–Italian government official with a deep appreciation for the arts. It was in large part shaped by an early education focused on the sciences and, in particular, his experiences accompanying the renowned and pioneering paleontologist and naturalist Louis Agassiz on his glacier expeditions in Switzerland.

Morelli brought the skills of an expert naturalist to the problem of looking at, comparing, and, finally, classifying works of art. In his major lifework, *Italian Painters: Critical Studies of Their Works*, Morelli foreshadows the still-unborn science of literary stylometry as he writes, "As most men, both speakers and writers, make use of habitual modes of expression, favorite words and sayings, which they often employ involuntarily and sometimes even most appropriately, so almost every painter has his own peculiarities, which escape him without being aware of it." According to Morelli, these "peculiarities" would find expression in the quiet corners of a work of art. Thus, the "Morellian method" relies on the comparison of seemingly minor details in paintings: folds in drapery, a fingernail, or an earlobe. In Morelli's view, it was only in details such as these, which he called the *Grundformen* (fundamental forms) of the artist, that the forces of tradition or schooling would be diminished enough so that the artist's true nature could shine through. In spirit, it is a visual form of Zipf's principle of least effort, a view of the artistic output as shaped by a battle between the expectations of the receiver and the predilections of the sender.

Morelli was able to apply his ideas with only the tools he had at hand, mainly his eyes. Lacking digital scanners and image-processing software, he built an internal database of *Grundformen* that, to some degree, would permit him to distinguish between details such as hands painted by Botticelli and Bellini. But with today's technology, we can begin to make mathematics out of Morelli. Through the use of *wavelet analysis*, a mathematical technique originally developed in the 1980s as a means of determining within sonar data the fingerprint of oil deposits beneath the ocean floor, my Dartmouth colleagues and I have made headway. We have found that a statistical summary of the density of simple linear elements in some of the drawings of the great Flemish artist Pieter Bruegel the Elder, as extracted from high-resolution digital scans of the originals, can provide a numerical signature that seems to act as a classifier for Bruegel's work. Even more Morellian than that, we have applied the same technique to the comparison of details within *Madonna and Child*, a huge altarpiece attributed to the great Renaissance master Perugino, in the hopes of determining the number of artists who contributed to the work. Whether or not the exploration for oil in the oceans has resulted in a definitive tool for the exploration of oils on the wall is still up for debate.

Wavelets are an example of a *multiscale analysis*—that is, they proceed by analyzing a work at successive levels of detail, like examining a work through varying the magnification on a microscope. In a wavelet analysis, what becomes important is the difference between what is measured at two scales, in a sense, representing the original image as some baseline structure to which is added a successive layering of detail.

Authentication through Fractals

Multiscale analysis is also at the heart of Richard Taylor's approach to Pollock. Taylor's investigations have not been driven by questions of provenance (although due to the frequent requests for authentication

PHASE II: UNIFIERS

he has set up a nonprofit company to manage this work as well as to protect himself in the case of lawsuits). He is mainly interested in the work, as both a scientist and an artist. He had taken a year's leave at one point in his career to devote himself to his own painting, but after a year decided that he was better off not quitting his day job. He brings a physicist's eye to the arts, and, in the case of Pollock, it has been a perfect storm of art and science that has enabled Taylor to find his own research in the works of this abstract expressionist master—the "chaos" for which one critic famously denounced Pollock's work in the 1950s is something that Taylor saw quite literally as the mathematics of fractal geometry.

The word *fractal* was coined in 1967 by the IBM mathematician Benoit Mandelbrot to encompass the geometric character of natural objects. The perfect lines, planes, and spheres of Euclidean geometry are Platonic abstractions, good for a first approximation to things like coastlines, landscapes, and clouds, but they clearly fall short at describing the variation of the natural world. Mandelbrot noticed that the character of such natural phenomena was a similarity in scale—that at each increase in magnification, the structures of nature, complicated though they are (*fractal* is derived from the same root as *fragment* and *fracture*), repeat themselves, maybe not precisely, but to a degree that can be quantified. The crags of a mountain range are replicated in the nooks and crannies of the stones that comprise them, or the eddies of a turbulent river flow are themselves composed of eddies within eddies within eddies. This is a piece of the connection between chaos and fractal geometry—the chaos that we now know colloquially in the metaphor of the hurricane in Texas generated by the flapping wings of the butterfly over China is a phenomenon that, when put into mathematical pictures (not unlike those that can be seen on The Weather Channel), yields images that exhibit this sort of self-similarity.

A famous example of a natural fractal is the irregular outline of

Chapter 14: The Numbers of Our Nature

the coast of England, which is, to a degree, replicated in any stretch of shore beneath the cliffs of Dover. The latter example, due to the British mathematician and polymath, Lewis Fry Richardson, was Mandelbrot's inspiration for the quantification of this irregularity in terms of its "fractal dimension," a number that effectively measures the complexity of a shape in terms of the degree to which it fills space at a given scale. The crinklier a line is, the more space it occupies in a box that surrounds it. Now, imagine a shape where, as you crank up the magnification, that sort of "misbehavior" is replicated: you've got yourself a true fractal.

> When Pollock so famously said, "I am nature," or "My concern is with the rhythm of nature . . . the way the ocean moves," he was possibly closer to the truth than anyone gave him credit for and probably closer than he knew himself.

A perfectly straight line has fractal dimension equal to one, while a square region has fractal dimension equal to two. Nature is generally somewhere in between: the coastline of England, the waves within waves of a stormy sea, the branches within branches of a fern leaf, our own circulatory or pulmonary system, or, as it turns out, the skeins of paint in a Pollock drip painting. That is Taylor's discovery. When Pollock so famously said, "I am nature," or "My concern is with the rhythm of nature . . . the way the ocean moves," he was possibly closer to the truth than anyone gave him credit for and probably closer than he knew himself.

Taylor has examined many of Pollock's works and found a remarkable degree of regularity in the fractal dimension that can be

PHASE II: UNIFIERS

computed by examining different color layers in the paintings. First, what is remarkable is that Pollock could regularly achieve fractal structure. Taylor's personal attempts at such a result were only successful when he came up with the idea of hanging a bedsheet from a tree and allowing the measurable fractal nature of the wind to be realized in dripped paint blown onto the sheet.

Even more, there appears to be a fractal dimension to Pollock's work that is characteristic of a given period, so that Pollock did, over periods of time, reliably reproduce in his work a small range of fractal dimensions. In fact, Taylor claims even more that in his examinations he finds evidence for two distinct fractal dimensions as might be predicted by a documented two-step working style in which Pollock would lay down a broad underlayer to which he would later add detail.

When presented with a would-be Pollock, Taylor performs the digital analysis and checks to see if the numbers jibe with those that have been computed for Pollock's known work of a given period. Taylor's analysis of the Matter collection suggested that the drip paintings were forgeries. However, in some related work, John Elton and Yang Wang of Georgia Tech; Jim Coddington, chief conservator at New York's Museum of Modern Art; and I have determined that a generalization of fractal dimension, called *multifractal analysis*, may provide a more textured signature for the work.

It's significant that Taylor found a digital signature for Pollock. But what might be even more significant is that the art world paid attention to it, for this shows the art–science boundaries are continuing to become fuzzier and fuzzier. Presumably, this is just the beginning, although there will surely be artists whose work defeats a statistical approach. In the spring of 2007, five teams of researchers will converge on the Van Gogh Museum in Amsterdam to present the results of a yearlong study aimed at uncovering a digital signature for Vincent van Gogh. And other methods of this type of work are emerging. An interesting and very general approach to finding a style in any digital media is work of the Dutch information theorist Paul Vitányi, whose analysis focuses on the information content (in a statistical sense) of the work. His media-free approach is one that allows any collection of numbers to be compared to any other, making possible the idea of comparing works of art to works of literature.

Stylometry opens us up to a world in which we are defined by our digital trail—the words we write, the websites we visit, the pictures we store, summarized in a statistical fingerprint. We are our actions. How very existential. 🌿

ON TIME AND RISK

Ole Peters, Imperial College London
SFI Bulletin, 2009

Let's say I offer you the following gamble: You roll a die, and if you throw a six, I will give you one hundred times your total wealth. Anything else, and you have to give me all that you own, including your retirement savings and your favorite pair of socks. I should point out that I am fantastically rich, and you needn't worry about my ability to pay up, even in these challenging times. Should you do it?

The rational answer seems to be "yes"—the expected return on your investment is 1,583⅓ percent in the time it takes to throw a die. But what's your gut feeling? Perhaps you are quite happy with your present situation; maybe you own a house and a nice car and a private jet—would you be one hundred times happier if you were one hundred times richer? And how much less happy would you be if you suddenly had nothing?

This example illustrates a common flaw in thinking about risky situations, one that can make us blind to excessive risks and which appears to have been a factor in the financial markets in recent years. As we will see, the calculation of the enormous expected return essentially assumes that you have dealings with parallel universes. Consequently, financial models can fall prey to the assumption that traders will regularly visit the parallel universe where everything comes up sixes. An analysis of risk and return that prohibits such eccentricities gives rather different answers. We will start with an outline of the classical treatment of risky problems, then offer an alternative, and finally discuss the practical consequences of both perspectives.

PHASE II: UNIFIERS

Daniel Bernoulli, the man who explained why helicopters fly a few hundred years after Leonardo da Vinci drew them and a few hundred years before they took to the skies, contemplated pretty much our gamble when, in 1738, he offered his answer to what economists now call the St. Petersburg paradox. The paradox asks how much a rational person should pay for a lottery ticket that offers a very low chance of a tremendous win.

He pointed out that mathematics alone does not capture the situation. It produces numbers for us like 1,583⅓ percent, but it cannot give those numbers meaning, for the fundamental reason that how much I own is irrelevant—what matters is what use my possessions are to me. I might require an expensive, life-saving operation next week, which limits my ability to take risky gambles. Or my name could be Diogenes, and when offered riches I yawn and mumble something about shade and sun, wave a hand, and turn around in my tubular abode. St. Exupéry's Little Prince comes to mind, who stares in bewilderment at the businessman who is counting the stars that he owns.

Bernoulli argued intuitively that the increase in the usefulness—utility—of my total wealth from a small gain should be inversely proportional to the wealth I already have. If I'm rich already, another dollar won't make much difference (although he also acknowledges exceptions, such as a rich man in prison whose utility increases more due to the extra ducats required to buy his freedom than that of a poorer man given the same amount). Mathematically expressed, this assumption amounts to a so-called *logarithmic utility function*. Utility functions had already been established before 1738 as a concept to reflect risk preferences and became the standard answer to problems where investments are characterized by an expected return and an uncertainty in that return.

Bernoulli's answer, logarithmic utility, reconciles the mathematics with our gut feeling—the expected utility (or logarithm) of your wealth after playing my game is negatively infinite, a strong warning against taking the gamble. But because his perspective is intuitive, it is vulnerable to modifications. Arguing on the basis of usefulness, different types of utility functions, designed to include rare exceptions like the rich prisoner, are no less valid than the logarithm he proposed. After all, these functions are supposed to reflect personal choices and circumstances. Thus, invoking the individuality of human beings, Bernoulli's peers emphasized that the full treatment of the problem is outside the realm of reason. But this sounds more like a cheap excuse than an answer to the problem—and, what's more, an excuse to choose a utility function that gives the answer I want.

A less vulnerable perspective that, strangely, remained on the fringes of economic theory was pointed out 218 years after Bernoulli's treatment of the problem by John Larry Kelly in 1956. I offer you the same bet as before. This time, following Kelly, we will make do without utility and instead focus on the irreversibility of time. Since we're considering a situation with randomness, we're interested in some expected, or average, performance. Playing the game repeatedly, we might expect the performance over many rounds to converge to this average.

Why might we expect this? If I ask you to roll your die one hundred times and tell me how many sixes you got, your answer will be somewhere around seventeen. Alternatively, we could measure the expected number of sixes by giving one die to each of one hundred people and let everyone roll once. In this instance, we will find a similar number of sixes—again, around seventeen. Whether we look at a time average (you rolling your die many times) or an ensemble average (many people each rolling a die once), as the number of trials increases the fractions of sixes will converge to $\frac{1}{6}$.

It seems trivial that the two differently computed averages should be the same—trivial enough for mathematical physicists

PHASE II: UNIFIERS

to question it. Ludwig Boltzmann, in about 1884, coined the term *ergodic* for situations with identical time averages and ensemble averages. Not every situation is like this, however; there exist "nonergodic" situations as well, and these are often as counterintuitive as the ergodic situations seem trivial.

So, do we have to be more careful when we talk about expected returns and average performances? There are two averages, not one—two ways of characterizing an investment, two quantities with different meanings. Let's consider each in turn, ask which one is relevant in our case, and see if they are identical.

First the ensemble average: when economists, or Bernoulli, speak of "expected return," they typically mean an average that is calculated as the sum over all possible outcomes, weighted by the probabilities of these outcomes. An example is the 1,583⅓ percent per round expected return of our game.

Probing a little deeper, we discover that this calculation uses the conceptual device of an ensemble of infinitely many identically prepared systems, or copies of our universe. The ensemble average simultaneously considers all possible paths along which the universe might evolve into the future. The fraction of systems from the ensemble that follows some scenario is the probability of that scenario, and summing the possible outcomes and weighted with their respective probabilities amounts to taking an average over all possible universes.

Herein lies the danger: if we don't actually play many identical games at once, then such an average only has practical relevance if it is identical to the quantity we're interested in, often the time average. There may be many possible paths from here into the future, but only one will be realized. In our game, you are risking your entire wealth, which obviously cannot be done many times simultaneously, so the ensemble average is not really the relevant

Chapter 15: On Time and Risk

quantity. Technically, it stems from a *gedankenexperiment* involving other universes.

Now the time average: perhaps it is identical to the ensemble average, and it doesn't matter which one we use. In other words, we ask, Is the situation ergodic? Considering the course of time, your ability to play the game tomorrow depends on the consequences of today's decisions, and next month's ability depends on the thirty daily outcomes in between. The ability of one player in the ensemble to play the game, on the other hand, does not depend on other players' luck. For this reason the ensemble average return is different from the time average—maliciously so: the time-average performance of a single investment is always worse than the ensemble average. So, unfortunately, the situation is not ergodic.

In our initial treatment of the game, the fact that I asked you to risk everything you own didn't impress the mathematics—it produced an expected return that seemed to strongly recommend playing the game. The reason this ensemble average didn't respond to the fact that you were most likely about to lose everything is this: the ensemble includes those few lucky copies of yourself whose enormous gains would easily make up for your likely loss.

Following Bernoulli, we reconciled the tempting expected return with our intuition by introducing utility. But this is not necessary—we simply need to recognize that we used an inappropriate average, implicitly treating the game as if we could interact with those parts of the ensemble that did not materialize (i.e., parallel universes) and realize the average return over all universes. If you find yourself in this situation, by all means, play the game. But if you're a mere mortal, I'd advise you not to do it. The time-average growth rate for this game, just like the expected logarithmic utility, is negatively infinite—if you don't believe me, play it a few times in a row. Instead of different changes in utility, the time perspective emphasizes that, as time goes by, we cut off different numbers of branches of potential universes reaching from the

PHASE II: UNIFIERS

present into the future. The difference in perspective is subtle but has far-reaching consequences.

We've considered an extremely risky game for illustration, but none of the above arguments are specific to it. In general, the time perspective reveals an upper limit on risks that may be considered sensible. For example, suppose I offered you a similar but different game: You get to roll a die and whatever you wager, I will give you one hundred times your wager if you throw a six. This situation is different because you can hold back some of your wealth in case you lose. In fact, the time perspective will tell you to invest about 16 percent of your net worth and keep playing the game, adjusting the wager to that same fraction after every round. It also tells you that over time you will realize a growth rate of about 33 percent per round. Crucially, *if you choose to risk more than this, you will gain less* (of course, you will also gain less if you risk less than 16 percent of your wealth).

A time-based approach provides insights into how to regulate credit rationally: how much an investment should be leveraged, the loan-to-value ratio at which a mortgage becomes a gamble, and the appropriate requirements for margins and minimum capital.

The literature on portfolio theory and risk management largely uses a combination of ensemble averages and utility, neglecting time or at best encapsulating its effects in a utility function. In this approach, time irreversibility, the unshakable physical motivation for refraining from excessive risk, is replaced by arbitrarily specifiable risk preferences. Following the establishment of the corresponding academic framework (roughly from the 1970s), regulatory constraints that were largely based on common sense were progressively loosened.

In an investment context, the difference between ensemble averages and time averages is often small. It becomes important, however, when risks increase, when correlation hinders diversification, when leverage pumps up fluctuations, when money is made cheap, when capital requirements are relaxed. If reward structures—such as bonuses that reward gains but don't punish losses, and also certain commission schemes—provide incentives for excessive risk, problems arise. This is especially true if the only limits to risk-taking derive from utility functions that express risk preference, instead of the objective argument of time irreversibility. In other words, using the ensemble average without sufficiently restrictive utility functions will lead to excessive risk-taking and eventual collapse. Sound familiar?

Considerations of time alone cannot capture an investor's or a society's risk preferences. These preferences will always depend on individual circumstances and include motivations, for example, moral motivations, that are indeed beyond the reach of mathematics. But time considerations do place objective upper bounds on advisable risks and go a long way toward rationalizing our intuitions.

Today's risk management often solely relies on investors specifying their risk preferences or, synonymously, their utility functions, without explicitly considering the effects of time. My bank asked me the other day what risk type I am, apparently expecting a reply like "I like a good gamble," or "I always wear my bicycle helmet." When I replied with a statement regarding time and answered, truthfully, that I'm the type who likes to see his money grow fast, they thought I was joking.

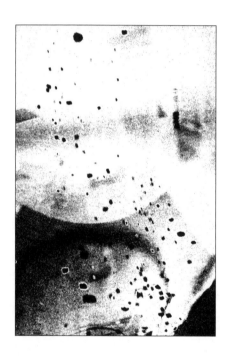

TRANSCIENCE: DISCIPLINES AND THE ADVANCE OF PLENARY KNOWLEDGE

David C. Krakauer, SFI
SFI Bulletin, 2011

"Everyone takes the limits of his own vision for the limits of the world."
—ARTHUR SCHOPENHAUER

Scientists, a risk-averse group, tend to eschew announcing their larger aims. After all, it is not entirely licit or proper to say, "We are trying to discern the laws of biology, or why social systems might proceed through sequences of increasing complexity," preferring instead remarks like "We are interested in gene regulation, or how large molecules are synthesized, or why the ancestral Puebloans stored maize." We feel that the larger objectives come across as grandiose, and so we retreat into prosaic descriptions of the work we do. In other words, we retreat into disciplinarity, a comfortable and familiar zone of tribal and historical cohesiveness, where the consolation of crowds helps to justify our activities. There is nothing wrong in cleaving to operational particulars, and for those interested in detail, these provide valuable information about what we do. The cost of this maneuver is that it restricts the scope of our inquiries and causes us to lose sight of the numerous extradisciplinary ideas and methods that have contributed to (and will be required to further) our progress through the thorny branches of science.

As we have systematically overcome our ignorance of the cosmos, we have pushed at the boundaries of natural phenomena, intermittently reaching critical points where the methods of a field have proven inadequate for further progress. New ideas, techniques, and devices imported from other fields have been required to squeeze through explanatory bottlenecks. Sometimes this fusion of fields has been

PHASE II: UNIFIERS

of sufficient magnitude to warrant the creation of a new discipline (genetics, ecology, etc.), and in time these absorb the insights of others. In this way, scientific disciplines possess something akin to a life cycle, with periods of rapid growth, maturation, sex, and finally senescence and even death. As the pace of life has accelerated, so has the production of disciplines and the rate of their extinction.

Scientists, as a professional order, were not recognized before 1837, when William Whewell coined the term in his *History of the Inductive Sciences*. As for science itself (excepting those who locate its true origins in the European scientific revolution of the seventeenth century), it is now widely accepted that scientific activities—meaning systematic forms of inductive–deductive process—have been ever present in human society. Best known of the pluralists is perhaps Joseph Needham, who in his *Science and Civilization in China* (first volume appeared in 1954, coauthored with Wang Ling) went to great lengths to demonstrate evidence of science and technology long before the European Renaissance, extending into the early millennia BC in China. These are efforts at locating concepts; however, we seek to explore their transmutation. It is not so much when and where science and scientists first appeared that interest us but the pace of scientific transformation. The geocentric model of the solar system proposed by Ptolemy in the *Alamagest* in the second century remained unchanged right up until the sixteenth century, when Copernicus proposed the more parsimonious heliocentric alternative. From Copernicus to Newton was just over a century, and from Newton to relativity, quantum mechanics, string theory, and dark matter, another couple of centuries.

The idea that all animals are preformed in the embryo (like nested matryoshka dolls) was the dominant theory of inheritance for most of our scientific history. Then, in 1865, the monk Gregor Mendel, while breeding peas, initiated the study of genetics. Genetics itself did not exist as a discipline outside of botany until

Chapter 16: Transcience: Disciplines and the Advance of Plenary Knowledge

William Bateson in 1894 coined the term in his *Materials for the Study of Variation*. At this point, the study of inheritance became a subject in its own right. In less than a century we have discovered DNA, regulatory RNA, prions, and the epigenome. Most of these are not studied in genetics departments (many of which were closed or renamed over the course of only a couple of recent decades, giving them a half-life of under a century) but rather in molecular biology, bioinformatics, and systems biology departments.

> Scientists, as a professional order, were not recognized before 1837, when William Whewell coined the term in his *History of the Inductive Sciences*.

The pattern we observe in the evolution of the scientific disciplines is what the late Buckminster Fuller characterized as accelerating acceleration, which implies that new ideas are appearing more quickly than we can possibly reorganize careers and departments to respond to them. The solution has been a messy mixed strategy, with new disciplines and journals popping up every year or month, and new ideas shoehorned into awkward groupings within existing departments to cope with the doctrinal flux. I am reminded of Oscar Wilde when he wrote, "Fashion is a form of ugliness so intolerable that we have to alter it every six months."

We have reached a stage where the pace of discovery and the nature of shared knowledge bring the whole venerable exercise of disciplinary fads into question. I believe we are entering a period of transcience, where it is becoming necessary that training in areas with fundamental mathematical, computational, and logical

PHASE II: UNIFIERS

principles should be emancipated from a single class of historically contingent case studies. For example, statistical physics will continue to be every bit as useful in understanding social phenomena as it traditionally has been in studying properties of condensed matter. The same could be said for suitable modification of computational theory and evolutionary dynamics. One of the significant contributions of SFI in this new landscape has been to show how ideas have a far greater compass than their original purpose suggests. Profound ideas are often characterized by considerable generality. Departments are becoming battlements that defend vested interests rather than idea incubators that advance understanding. Transcience is an expression that seeks to recognize the pursuit of plenary or synthetic knowledge as an institutional priority.

There are those who would argue that, without the rigors of traditional disciplinary instruction, we shall be producing researchers capable of little more than shallow metaphor construction. By their reckoning, the correct approach to complex phenomena is to first apprentice ourselves to tried-and-true research projects. This is the familiar "when I was a lad I got up at four a.m. and walked fifteen miles to work" line of reasoning. The alternative is not to neglect the details of a system but to recognize that many of our most pressing problems and most interesting challenges reside at the boundaries of existing disciplines and require the development of an entirely new kind of sensibility that remains "disciplined" by careful empirical experiment, observation, and analysis. We are not losing depth but are recognizing the full potential of theoretical frameworks of significant universality, and that these should not be limited to communities based on their historical development. Ours is a landscape that can support diversity, and those with disciplinary separation anxiety are free to persist as they are.

The sciences of complexity are our best working examples of transcientific research, but they remain restricted in part through

Chapter 16: Transcience: Disciplines and the Advance of Plenary Knowledge

the association of complexity with a small class of models. In this issue of the *Bulletin*, we observe the continued maturation of the field of complexity as we accrue more data, hone our intuitions, and extend the scope of our theories. From the study of cities, through conflict, technological innovation, and cognition, we find a multitude of shared patterns amenable to overlapping forms of analysis. This issue is not organized into sociology, biology, engineering, and neuroscience—none of which would provide an adequate classification for the work being described. Readers of the *Bulletin* are fully aware that each of these areas of inquiry will obdurately resist shoehorning into a disciplinary framework, and there is absolutely no good reason to try. Perhaps it is time for our schools, universities, and research institutes to embrace the full implications of this shift in thought and to redesign curricula and perhaps even demolish a few departments accordingly. We are entering a phase of increasingly transcientific research, and it is time society and academia wake up to the full implications of this reality.

WHAT BIOLOGY CAN TEACH US ABOUT BANKING

Lord Robert May, Oxford University
SFI Bulletin, 2012

The behavior of nonlinear dynamical systems has been the unifying theme of my own nonlinear academic trajectory. Beginning as an undergraduate chemical engineer, I ended up with a PhD in theoretical physics, and roughly ten years later transmogrified into a professor of biology at Princeton University. I believe the ways in which system risks can arise, and propagate, in different settings is best seen from many different perspectives. And it is increasingly clear that such a view of complex adaptive systems is critical to our future well-being, as we are indeed engulfed in complex, and often coupled, systems, from our environment to our social networks and our financial systems.

In my own subject of ecology, SFI has been a major player in understanding systemic risk, particularly in studies of the nonrandom network structures whereby real-world ecosystems reconcile complexity (many species interacting with each other) with persistence in naturally fluctuating environments. Given the additional shocks being imposed on ecological systems by human activities—overexploitation, habitat destruction, alien introductions, all compounded by climate change—such understanding is increasingly important. It is especially so as we strive to maintain a multitude of ecosystem services, not counted in conventional assessment of gross domestic product but upon which we depend. In this general area, SFI professors such as Jennifer Dunne, Mercedes Pascual, and others are among the best in the business.

This is only one of several major areas where SFI's "clean sheet of paper" approach to complicated problems has been important. It is my belief, however, that the recent—and continuing—worries

PHASE II: UNIFIERS

about the performance of financial markets present SFI with its greatest-yet challenge and concomitant opportunity.

Figure 1 provides a striking illustration of the truly extraordinary growth in the amount of leveraged money swishing around within the UK banking system in recent years, arguably associated in part with the growth of computing power and contrasting greatly with the previous century's stability. Other countries show similar patterns. Much of this growth derives from increasingly complex financial instruments, which purport to reconcile greater returns with diminished risks.

In 2006, the US National Academy of Sciences (NAS) and the Federal Reserve Bank of New York (FRBNY) put together a prescient study, based on the observation that, while such complex "derivatives" and credit-default swaps seemed attractive at the level of individual financial institutions (henceforth brigaded as "banks"), essentially no one was considering the possible implications for the system as a whole. In addition to bankers and other economists, this NAS/FRBNY study

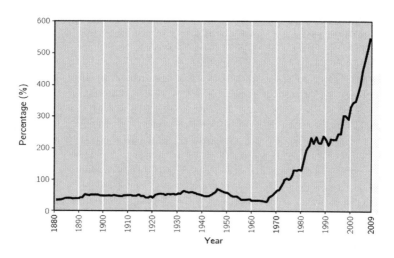

FIGURE 1. This graph illustrates UK bank assets expressed as a percentage of GDP, over the past century. The fast increase may be due to the amount of leveraged money circulating within the system. CHART RE-CREATION: JAN UNDERWOOD/ INFOBURRITO.COM

Chapter 17: What Biology Can Teach Us About Banking

drew in researchers from areas where some "read-across" seemed likely: ecology, infectious disease transmission, and the electricity grid.

Subsequent to the financial crisis that began in 2008, this issue of systemic risk has moved center stage. In the UK, studies of mathematical metaphors or "toy models" of banking systems have buttressed the intuition of central bankers, suggesting, for example, that all banks should revert to the somewhat higher capital reserves (or other liquidity) that they had previously held. These studies also suggest that big banks should hold relatively bigger such reserves than small banks (contrary to the trends of the nineties and noughties); there are lessons to be learned about the disproportionate influence of big banks from relatively recent work on "super-spreaders" of infectious diseases.

Additionally, the stabilizing advantages of modular organization in complex systems, seen both empirically and theoretically in ecosystems, suggests a return to greater separation between retail and investment banking activities, along the lines of the US Glass–Steagall Act enacted in 1933. This legislation followed the recognition that a major factor in the Great Depression was banks leveraging casino activity with depositors' money. (Glass–Steagall was repealed at the high tide of free-market extremism that flourished toward the twentieth century's end.) These measures, and broadly similar ones being aired in the US, not only march with the dynamical properties of sensible models of banking systems but also are intuitively reasonable.

The recommendations of the UK's Independent Banking Commission, reported on September 12, 2011, are broadly along the above lines: in particular, higher capital reserves and retail banking activities to be structurally separated (by a strong but flexible "ring fence") from wholesale and investment activity. Many bankers, however, argue against these recommendations and simply wish to get back on the roller coaster.

PHASE II: UNIFIERS

All these problems are compounded by the fact that there can be a conflict between what is best for any one bank viewed in isolation and what is best for the system. This paradox is exemplified by the following toy model (fig. 2): Consider N banks and N distinct, uncorrelated asset classes, each of which has some very small probability, p, of having its value decline to the extent that a bank holding solely that asset would fail. At one extreme, assume each bank holds the entirety of one of the N assets: the probability for any one bank to fail is now p, whereas that for the system is a vastly smaller p^N. At the opposite extreme, assume all banks are identical, each holding $1/N$ of every one of the N assets: the probability for any one bank to fail is now much smaller than p, but all banks now being identically constituted, if one fails, all fail, and this probability is much bigger than p^N (being of the general order $e^N p^N$). The former pattern minimizes diversification of individual banks but maximizes diversity of the system, whereas the latter does the opposite. Previous international banking regulatory meetings, known as Basel I and II, had focused on individual banks and essentially disregarded the system

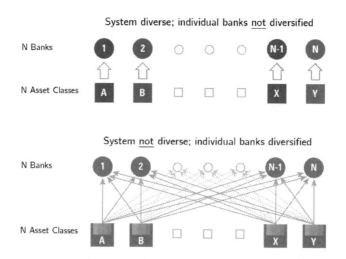

FIGURE 2. The top arrangement minimizes diversification of individual banks but maximizes diversity of the system, whereas the bottom arrangement does the opposite. CHART RE-CREATION: JAN UNDERWOOD/INFOBURRITO.COM

Chapter 17: What Biology Can Teach Us About Banking

as a whole. Such promotion of individual diversification, without corresponding attention to systemic risk, thus arguably contributed to our present problems.

My view is that considerations of systemic risk are very important, and that greater understanding of how to minimize the likely costs of problems cascading through the system is needed. But I also believe it to be of even greater importance to have more sophisticated and reliable mechanisms for rating complex financial instruments. In retrospect, it is hard to believe anyone could have been so bewitched by illusionary mathematical elaboration of faulty assumptions as to rate collections of Triple B house mortgages as Triple A. There are both technical and social questions here: not only was the mathematics underpinning the evaluation of complex derivatives (arbitrage pricing theory) grossly unsound, but excessively diligent credit ratings agencies are unlikely to survive in a privatized system. How best to resolve this problem?

Underlying the problems of systemic risk and of proper evaluation of individual financial instruments is a deeper and even more difficult question, recently posed by Harvard's distinguished economist Benjamin Friedman. Beginning with the observation that the role of financial markets in a free-enterprise economy is the efficient allocation of investment capital, he went on to ask, "How much is it costing us to operate this financial system?" His answer: "A lot." Quantifying this assessment, he observed that, in the US, thirty years ago the cost of running the financial system was 10 percent of all the profits earned in America. This rose to 20–25 percent fifteen years ago, and just before the crisis hit, "running the financial system took one-third of all profits earned on investment capital."

I thus conclude by suggesting one important and appropriate task for the sciences of complexity, and for SFI. Take up Benjamin Friedman's challenge: "The time has come for a serious evaluation of the costs and benefits of running our financial system."

IMAGINING COMPLEX SOCIETIES

Scott G. Ortman, SFI
SFI Bulletin, 2013

The societies in which most humans live have changed dramatically over the past 10,000 years. At the end of the last ice age, all humans lived in hunting and gathering bands where nearly everyone was related, generalized reciprocity was the norm, families produced almost everything they needed, and group decision-making was consensual. In contrast, today most of us live in industrial nation-states where we will never meet most of our compatriots, economic exchange is the norm, families produce only the tiniest fraction of the goods and services they need, and political decisions are made through bureaucratic governments.

These changes make clear that the complexity of many human societies—as defined by their scale, functional differentiation, and control structures—has increased dramatically in recent millennia. How and why this occurred is one of the central questions of anthropology, but despite sustained attention we are still a long way from a truly scientific understanding. In this essay I'll offer my own view of the problem and what I think is needed to move research in this area forward.

My point of departure is Bruce Trigger's *Understanding Early Civilizations* (2003), the most detailed comparative analysis of early state societies yet produced. Trigger chose to work with a sample of early civilizations that developed independently, had never been subservient to other societies, and for which both archaeological and written sources are available. As a result, he did not compare the earliest state societies in various world regions but compared the earliest ones for which a well-rounded picture is possible. Thus, he

PHASE II: UNIFIERS

examined the Aztec (1100–1519 CE) as opposed to the Teotihuacan (100 BCE–750 CE) period in Central Mexico, the Early Dynastic III (2500–2350 BCE) as opposed to the Uruk (3400–3100 BCE) period in Mesopotamia, and so forth. In his view, the disadvantages of working with civilizations from more recent periods were outweighed by the advantages of examining the richer available evidence for the symbolic and cognitive aspects of each society, in addition to their economic and sociopolitical structures.

His basic findings are striking. First, the economies of early civilizations were highly variable and reflected the process of local adaptation to the specific environments in which each emerged. Second, the political organizations of early civilizations also varied, falling into one of two basic types: city-states, where a large number of farmers lived with elites in urban centers and full-time craft specialists produced goods that were distributed to all through markets; and territorial states, where most farmers lived outside of urban centers and craft specialists produced goods primarily for the elite. Third, the religious beliefs of early civilizations did *not* vary. In all these societies, relations between humans and the forces of nature were imagined as parallel to relations between commoners and rulers. Anthropomorphized forces of nature required material sacrifices in order to persist and fulfill their roles in maintaining the natural order; and, in the same way, elites required surpluses and labor from commoners in order to fulfill their roles in maintaining the social order.

To the extent that variation across independent cases implies latitude in adaptive possibilities, and uniformity implies constraints, these findings imply that the strongest constraint in the emergence of early civilizations was beliefs that supported new scales of social coordination. This is in strong contrast to the view, enshrined in many approaches to human behavior, that the primary constraints were material or technological. Trigger's results suggest instead that the way forward in our efforts to understand social complexity is to focus on the process by which beliefs that support complexity were invented and adopted within societies. In other words, we need a better understanding of how human society itself emerges from shared abstract ideas, or what I would call cultural models. When viewed from this perspective, many traditional explanations for the emergence of complex societies turn out to depend on and presuppose this more fundamental process. It is clear that agricultural intensification, economic and bureaucratic specialization, technological advances, and warfare were all involved, but what is it that makes people feel it is safe to invest in farmland or to depend on others for the goods and services they need; that it is natural to hand over surpluses to rulers; or that it is appropriate to kill people who have never harmed them directly? Trigger's results suggest human imagination was much more central to this process than we have previously considered.

Articulating how abstract ideas that promote social coordination are invented and spread through society is a challenging task, but due to progress in several fields it is becoming possible to sketch an outline of how it might occur. The first point to recognize is that economies of scale are intrinsic properties of human social networks. This has been amply demonstrated for contemporary urban systems (Bettencourt et al. 2007; Bettencourt et al. 2010), and it is also apparent in the archaeological record. For example, figure 1 plots the population vs. the settled area of the largest site in archaeological traditions from around the globe. The power law fit to these data, which span five orders of magnitude, exhibits the precise economy of scale, in the form of area per person, predicted by urban scaling theory (see Bettencourt 2012); but, in this case, each point represents a settlement that developed in a different cultural tradition, with a different technological and economic base, and in a different part of the world. These data make a strong case that, as

PHASE II: UNIFIERS

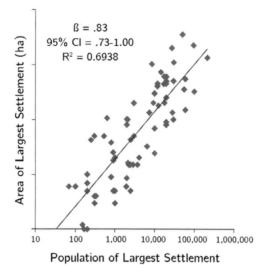

FIGURE 1. Population vs. settled area of the largest settlements in archaeological traditions from around the world. The fit line is a power law with exponent β. Data from Ortman and Blair 2012

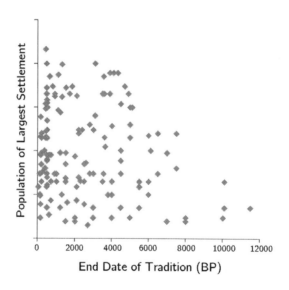

FIGURE 2. End dates and largest settlement populations of archaeological traditions from around the world. Data from Ortman and Blair 2012

human social networks grow, they necessarily lead to systems that require fewer resources per person, and produce more per person. In other words, the benefits of scale for human groups have always been there.

If this were all there is to the problem of social complexity, one might expect all human societies to have grown consistently in scale and complexity over time, but this is not what has happened. Figure 2, for example, plots the age of various archaeological traditions against the population of the largest settlement in each tradition (a reasonable proxy for overall complexity, see Chick 1997; Naroll 1956). The chart shows that the range of complexity in human societies has grown exponentially since the end of the last ice age, but many societies have remained simple over this period. It's also important to emphasize that societies where complexity accumulated were not always located in more productive areas, or in areas where agriculture had been practiced the longest (Ortman and Blair 2012). So why has complexity accumulated only in certain societies, despite the intrinsic benefits of scale?

The answer derives from the fact that what is good for groups is not always good for the individuals comprising them. For example, both multicellular organisms and social insect colonies are functionally specialized and hierarchically organized collectives that are highly successful in maintaining and transmitting accumulated knowledge, in the form of genetic instructions, to the next generation; but they also have little regard for the fates of most cells or insects (Maynard Smith and Szathmáry 1999). This same pattern is apparent, in an attenuated way, in human societies. For example, economist George Steckel and anthropologist Jerome Rose (2002) examined health indicators for Prehispanic New World societies and found that the median health of individuals declined as societies grew more complex. This suggests social complexity emerges from mechanisms that promote coordinated behavior even if it is not in the best interest of each individual. In the case of multicelled organisms and insect colonies, the solution was to make the coordinating individuals (cells, insects) genetic clones or siblings. That way, genes that promote cooperation could spread even if the most cooperative individuals left no offspring. What was the solution for humans?

I think the solution has a two-part answer. First, humans do possess some groupish predispositions that have evolved since the genus *Homo* became distinct from our living ape relatives. Psychologist Jonathan Haidt (2012) argues that these moral instincts—especially concerns over care, fairness, loyalty, authority, and sanctity—coevolved with the human capacity for language and culture. Economists Sam Bowles and Herb Gintis (2011) have also shown that the conditions faced by early humans were appropriate for the evolution of these predispositions. As a result, it appears reasonable to conclude that the characteristics of early small-scale human societies reflect the mix of selfish and groupish instincts characteristic of human nature. If so, the subsequent accumulation of social complexity in some societies would seem to derive from the ways particular cultural models, invented by particular humans in specific contexts, interfaced with human moral and emotional instincts.

Given this first part of the answer, the second part derives from neuroscientist Antonio Damasio's (1994) model of human decision-making. His model has the following elements. First, humans possess evolved hormonal and neural circuits that induce responses to various stimuli automatically. Think of what happens when you touch a hot stove, get thirsty or hungry, or are startled by a sudden movement or sound. The cascades of responses, including those related to our moral instincts, are known as primary emotions. Second, our nervous system continuously monitors our body state and stores "images" of the body states associated with our

Chapter 18: Imagining Complex Societies

experiences. If you've ever gone for a hike without water, "images" of the resultant thirst and anxiety become part of your memory of the experience. These images of body-state responses are known as secondary emotions. Third, humans form intuitions regarding potential courses of action through "gut feelings," which is to say, by associating specific instances with models of our previous experience, including their associated secondary emotions. As a result, human preferences typically derive from the implied emotional outcomes of alternative courses of action.

Damasio's model gets us part of the way to an explanation for coordinated behavior, but there is one final, crucial step where human imagination takes center stage, recruiting human nature in the service of social goals. The key insight comes from the linguist George Lakoff and philosopher Mark Johnson (1980), who demonstrated that abstract human thought is fundamentally metaphorical: we typically use the imagery of relatively concrete domains of experience to form intuitions about more abstract domains, especially in the social, political, and ecological realms. Most important, the source domains of these conceptual metaphors ultimately derive from our basic bodily experience, including associated secondary emotions. For example, contemporary Americans often imagine a nation as a body in forming opinions about social policies, and psychological experiments show that body-state imagery influences this process (Landau et al. 2009). Also, in my own research on Tewa Pueblo origins, I've found that imagining the community as a garden, with women as corn plants and men as clouds, was central to the emergence of an intercommunity ceremonial system that supported permanent villages and community-level specialization (Ortman 2012). These observations suggest that social complexity ultimately emerges from people behaving in terms of the body-state imagery of their shared social metaphors. (If you habitually imagine your community as a family, and you have experienced loving parents, then surely your leaders have your best interests at heart.) In social insects, chemical circuits encoded by genes induce coordinated behavior automatically; in humans, culture achieves similar results by linking models of the social, political, economic, and ecological worlds to our automatic and evolutionarily ancient emotional response systems, including moral instincts. And the more deeply ingrained these metaphors are, the more effectively they channel human behavior.

This is not to deny that human societies maintain competing models of the social world, that some individuals behave in accordance with dominant models simply because it is the path of least resistance, or that others actively resist these norms. There is also still a lot to learn about why specific metaphors are persuasive in certain contexts and not in others. Humans are not ants. Nevertheless, deeper reflection on the role of human imagination reveals its fundamental role. Put simply, the earth could not support as many people as it does today if humans had not invented the concept of government from our basic experiences of family life, or the concept of money from our experiences trading small and precious objects. All good ideas seem obvious once someone has them, but the cultural models that subsidized the accumulation of social complexity, and which seem natural to us today, were not self-evident to our distant ancestors. Instead, these models had to be invented and promoted. Once invented, they could spread for a variety of reasons, but they didn't have to. As in biology, I suspect that both material and cultural (a.k.a. political) constraints—cultural genotypes, if you will—influenced the process of invention and adoption, and there was significant path dependence (Wagner 2011).

These details aside, the research reviewed here suggests that a fundamental factor in the emergence of complex societies was new cultural models that recruited the emotional concerns and

Chapter 18: Imagining Complex Societies

moral instincts of farmers and herders in support of hierarchical and functionally specialized organization. For example, in Uruk, Mesopotamia, the world's first city-states were founded on the idea of the king as the good shepherd: the king protected and provided for his human flock, and it was thus right and natural for his subjects to obey him (Algaze 2008, 128–29). Many readers will recognize that this imagery continues to play a role in all three of the world's major monotheistic religions. In other parts of the world, the specific imagery was different (among ancient Maya people, for example, the king was maize), but in all cases these cultural models emphasized the benefits of hierarchical and functionally specialized organization while hiding its disadvantages, thus tipping the scales of moral intuitions and public sentiment in favor of larger-scale social coordination. And there is little doubt that these societies have been spreading ever since, for better and for worse.

> The earth could not support as many people as it does today if humans had not invented the concept of government from our basic experiences of family life, or the concept of money from our experiences trading small and precious objects.

At this point, the outline sketched here is little more than a qualitative framework. Much work remains to be done to translate this framework into a quantitative and testable model and to assess the influence of cultural models in comparison with other factors that clearly were involved in the emergence of complex human societies. This will take time, hard work, and good collaborators. But if we are ever to understand the fundamental nature

PHASE II: UNIFIERS

of human societies and why they seem to be becoming more complex all the time, I believe this is the direction in which we should be working. ✤

REFERENCES CITED

Algaze, G. 2008. *Ancient Mesopotamia at the Dawn of Civilization.* Chicago: University of Chicago Press.

Bettencourt, L. M. A. 2012. "The Origins of Scaling in Cities." Santa Fe Institute Working Papers 12-09-014.

Bettencourt, L. M. A., J. Lobo, D. Helbing, C. Kühnert, and G. B. West. 2007. "Growth, Innovation, Scaling, and the Pace of Life of Cities." *Proceedings of the National Academy of Sciences of the USA* 104: 7301–6.

Bettencourt, L. M. A., J. Lobo, D. Strumsky, and G. B. West. 2010. "Urban Scaling and Its Deviations: Revealing the Structure of Wealth, Innovation and Crime across Cities." *PLoS ONE* 5(11): e13541.

Bowles, S., and H. Gintis. 2011. *A Cooperative Species: Human Reciprocity and Its Evolution.* Princeton, NJ: Princeton University Press.

Chick, G. 1997. "Cultural Complexity: The Concept and Its Measurement." *Cross-Cultural Research* 31(4): 275–307.

Damasio, A. 1994. *Descartes' Error: Emotion, Reason, and the Human Brain.* New York: G. P. Putnam.

Haidt, J. 2012. *The Righteous Mind: Why Good People Are Divided by Politics and Religion.* New York, NY: Pantheon Books.

Lakoff, G., and M. Johnson. 1980. *Metaphors We Live By.* Chicago, IL: University of Chicago Press.

Landau, M. J., D. Sullivan, and J. Greenberg. 2009. "Evidence That Self-Relevant Motives and Metaphoric Framing Interact to Influence Political and Social Attitudes." *Psychological Science* 20(11): 1421–27.

Maynard Smith, J., and E. Szathmary. 1999. *The Origins of Life: From the Birth of Life to the Origins of Language.* Oxford, UK: Oxford University Press.

Naroll, R. 1956. "A Preliminary Index of Social Development." *American Anthropologist* 56: 687–715.

Ortman, S. G. 2012. *Winds from the North: Tewa Origins and Historical Anthropology.* Salt Lake City, UT: University of Utah Press.

Ortman, S. G., and L. Blair (2012) "Expanding the Atlas of Cultural Evolution." Presentation at the Workshop "The Principles of Complexity: Life, Scale, and Civilization," August 6–8, 2012, Santa Fe Institute, Santa Fe, New Mexico.

Steckel, R. H., and J. C. Rose, eds. 2002. *The Backbone of History: Health and Nutrition in the Western Hemisphere*. Cambridge, UK: Cambridge University Press.

Trigger, B. G. 2003. *Understanding Early Civilizations*. Cambridge, UK: Cambridge University Press.

Wagner, A. 2011. *The Origins of Evolutionary Innovations*. Oxford, UK: Oxford University Press.

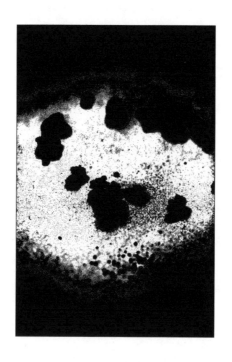

COMPLEXITY: A DIFFERENT WAY TO LOOK AT THE ECONOMY

W. Brian Arthur, SFI and Palo Alto Research Center
SFI Bulletin, 2014, Issue 2

Economics is a stately subject, one that has altered little since its modern foundations were laid in Victorian times. Now it is changing radically. Standard economics is suddenly being challenged by a number of new approaches: behavioral economics, neuroeconomics, new institutional economics. One of the new approaches came to life at the Santa Fe Institute: complexity economics.

Complexity economics got its start in 1987 when a now-famous conference of scientists and economists convened by physicist Philip Anderson and economist Kenneth Arrow met to discuss the economy as an evolving complex system. That conference gave birth a year later to the Institute's first research program—the Economy as an Evolving Complex System—and I was asked to lead this. That program in turn has gone on to lay down a new and different way to look at the economy.

To see how complexity economics works, think of the agents in the economy—consumers, firms, banks, investors—as buying and selling, producing, strategizing, and forecasting. From all this behavior markets form, prices form, trading patterns form: aggregate patterns form. Complexity economics asks how individual behaviors in a situation might react to the pattern they together create, and how that pattern would alter itself as a result, causing the agents to react anew.

This is a difficult question, so, traditionally, economics has taken up a simpler one. Conventional economics asks how agents' behaviors (actions, strategies, forecasts) would be upheld by—would be

PHASE II: UNIFIERS

consistent with—the aggregate patterns these cause. It asks, in other words, what patterns would call for no changes in microbehavior, and would therefore be in stasis or equilibrium.

The standard equilibrium approach has been highly successful. It sees the economy as perfect, rational, and machinelike, and many economists—I'm certainly one—admire its power and elegance. But these qualities come at a price. By its very definition, equilibrium filters out exploration, creation, transitory phenomena: anything in the economy that takes adjustment—adaptation, innovation, structural change, history itself. These must be bypassed or dropped from the theory.

By the mid-1980s, many economists were ready for a change.

Just what that change would consist of we were not quite sure when our program began. We knew we wanted to create an economics where agents could react to the outcomes they created, where the economy was always forming and evolving and not necessarily in equilibrium. But we didn't quite know how to achieve that.

In fact, in 1988 the Institute was still very much a startup. The program consisted in its first two years of twenty or so people, several of whom proved central: John Holland, Stuart Kauffman, David Lane, and Richard Palmer. We would meet, in an early version of what became Santa Fe Institute style, in the kitchen of the old convent on Canyon Road in the late mornings and loosely discuss ways forward.

These "emerged" slowly—sometimes painfully—mainly by talking over why economics did things the way it did and how alternatives might work. Our group was motley, even eccentric. Halfway through the first year the journalist James Gleick asked me how I would describe my group. I was hard put to reply. He pressed the question. Finally I said, "You remember the bar in *Star Wars*, at the end of the galaxy with all the weird creatures, Chewbacca and the others? That's our group."

Chapter 19: A Different Way to Look at the Economy

We did have some tools. We had new stochastic dynamic methods, and nonlinear dynamics, and novel ideas from cognitive science. And of course we had computers. But it took us a couple of years before we realized we were developing an economics based not just on different methods but on different assumptions.

> We found ourselves creating "artificial worlds"—miniature economies within the computer—where the many players would be represented by little computer programs that could explore, respond to the situation they together created, and get smarter over time.

Instead of seeing agents in the economy as facing perfect, well-defined problems, we allowed that they might not know what situation they were in and would have to make sense of it. Instead of assuming agents were perfectly rational, we allowed there were limits to how smart they were. Instead of assuming the economy displayed diminishing returns (negative feedbacks), we allowed that it might also contain increasing returns (positive feedbacks). Instead of assuming the economy was a mechanistic system operating at equilibrium, we saw it as an ecology—of actions, strategies, and beliefs competing for survival—perpetually changing as new behaviors were discovered.

Other economists—in fact, some of the greats like Joseph Schumpeter—had looked at some of these different assumptions before, but usually at one assumption at a time. We wanted to use

PHASE II: UNIFIERS

all these assumptions together in a consistent way. And other complexity groups in Brussels, France, Ann Arbor, and MIT were certainly experimenting with problems in economics. But we had the advantage of an interdisciplinary critical mass for a program that ran across all of economics. The result was an approach that saw economic issues as playing out in a system that was realistic, organic, and always evolving.

Sometimes we could reduce the problems we were studying to a simple set of equations. But just as often our more challenging assumptions forced us to study them by computation. We found ourselves creating "artificial worlds"—miniature economies within the computer—where the many players would be represented by little computer programs that could explore, respond to the situation they together created, and get smarter over time.

Our artificial-worlds-in-the-computer approach, along with the work of others both inside and outside economics, in the early 1990s became agent-based modeling, now a much-used method in all the social sciences.

One early computer study we did was a model of the stock market. In a stock market, investors create forecasts from the available information, make bids and offers based on these, and the stock's price adjusts accordingly. Conventional theory assumes homogeneous investors who all use identical forecasts (so-called rational expectations) that are consistent with—on average validated by—the prices these forecasts bring about. This gives an elegant theory, but it begs the question of where the identical forecasts come from. And it rules out transitory phenomena seen in real markets, such as bubbles and crashes and periods of quiescence followed by volatility.

We decided to have "artificial investors" in our computer create their own individual forecasts. They would start with random ones, learn which worked, form new ones from these, and drop poorly performing ones. Forecasts would thus "compete" in a mutually created ecology of forecasts. The question was, How would such a market work? Would it duplicate the standard theory? Would it show anything different?

When we ran our computerized market, we did see outcomes similar to those produced by the standard theory. But we saw other phenomena, ones that appeared in real markets. Some randomly created forecasts might predict upward price movement if previous prices were trending up; other types of forecasts might foretell a price fall if the current price became too high. So if a chance upward movement appeared, the first type would cause investors to buy in, causing a price rise and becoming self-affirming. But once the price got too high, the second sort of forecast would kick in and cause a reversal. The result was bubbles and crashes appearing randomly and lasting temporarily.

Similarly, periods of quiescence and volatility spontaneously emerged. Our investors were continually exploring for better forecasts. Most of the time this created small perturbations.

But occasionally some would find forecasts that would change their behavior enough to perturb the overall price pattern, causing other investors to change their forecasts to re-adapt. Cascades of mutual adjustment would then ripple through the system. The result was periods of tranquility followed randomly by periods of spontaneously generated perturbation—quiescence and volatility.

The program, as it developed, studied many other questions: the workings of double-auction markets; the dynamics of high-tech markets; endogenously created networks of interaction; inductive reasoning in the economy. In an SFI program parallel to ours, Josh Epstein and Rob Axtell created an artificial society called Sugarscape in which cooperation, norms, and other social phenomena spontaneously emerged. And in 1995 John Miller and Scott Page started an annual workshop in computational social sciences at SFI where

PHASE II: UNIFIERS

postdocs and graduate students could get practical training in the new methods.

The approach finally received a label in 1999, when an editor at *Science* asked me on the phone to give it a name. I suggested "complexity economics," and that name stuck.

Things have widened a great deal since then. Doyne Farmer has taken up studies of how technologies improve over time. And he, Axtell, and others have been using large datasets, along with agent-based modeling methods, to understand the recent housing-market crisis. Other groups in the United States and Europe have been using complexity methods to look at economic development, public policy, international trade, and economic geography.

None of this means the new, nonequilibrium approach has been easily accepted into economics. The field's mainstream has been interested but wary of it. This changed in 2009 after the financial meltdown when, as the *Economist* magazine observed dryly, the financial system wasn't the only thing that collapsed; standard economics had collapsed with it. Something different was needed, and the complexity approach suddenly looked much more relevant.

Where does complexity economics find itself now? Certainly, many commentators see it as steadily moving toward the center of economics. And there's a recognition that it is more than a new set of methods or theories: it is a different way to see the economy. It views the economy not as machinelike, perfectly rational, and essentially static, but as organic, always exploring, and always evolving—always constructing itself.

Some people claim that this economics is a special case of equilibrium economics, but actually the reverse is true. Equilibrium economics is a special case of nonequilibrium and, hence, of complexity economics.

Complexity economics is economics done in a more general way. In 1996, a historian of economic thought, David Colander, captured the two different outlooks in economics in an allegory. Economists, he says, a century ago stood at the base of two mountains whose peaks were hidden in the clouds. They wanted to climb the higher peak and had to choose one of the two. They chose the mountain that was well defined and had mathematical order, only to see when they had worked their way up and finally got above the clouds that the other mountain, the one of process and organicism, was far higher. Many other economists besides our Santa Fe group have started to climb that other mountain in the last few years. There is much to discover.

LIFE'S INFORMATION HIERARCHY

Jessica C. Flack, University of Wisconsin–Madison
SFI Bulletin, April 2014

Biological systems—from cells to tissues to individuals to societies—are hierarchically organized (e.g., Feldman and Eschel 1982; Buss 1987; Smith and Szathmáry 1998; Valentine and May 1996; Michod 2000; Frank 2003). To many, *hierarchical organization* suggests the nesting of components or individuals into groups, with these groups aggregating into yet larger groups. But this view—at least superficially—privileges space and matter over time and information. Many types of neural coding, for example, require averaging or summing over neural firing rates. The neurons' spatial location—that they are in proximity—is, of course, important, but at least as important to the encoding is their behavior in time. Likewise, in some monkey societies, as I will discuss in detail later in this review, individuals estimate the future cost of social interaction by encoding the average outcome of past interactions in special signals and then summing over these signals.

In both examples, information from events distributed in time as well as space (fig. 1) is captured with encodings that are used to control some behavioral output. My collaborators and I in the Center for Complexity & Collective Computation (C4) are exploring the idea that hierarchical organization at its core is a nesting of these kinds of functional encodings. As I will explain, we think these functional encodings result from biological systems manipulating space and time (fig. 2) to facilitate information extraction, which in turn facilitates more efficient extraction of energy.

This information hierarchy appears to be a universal property of biological systems and may be the key to one of life's greatest

PHASE II: UNIFIERS

mysteries—the origins of biological complexity. In this essay, I review a body of work by David Krakauer, myself, and our research group that has been inspired by many years of work at the Santa Fe Institute (e.g., Crutchfield 1994; Gell-Mann 1996; Gell-Mann and Lloyd 1996; Fontana and Buss 1996; West, Brown, and Enquist 1997; Fontana and Schuster 1998; Ancel and Fontana 2000; Stadler, Stadler, Wagner, and Fontana 2001; Smith 2003; Crutchfield and Görnerup 2006; Smith 2008). Our work suggests that complexity and the multiscale structure of biological systems are the predictable outcome of evolutionary dynamics driven by uncertainty minimization (Krakauer 2011; Flack 2012; Flack, Erwin, Elliot, and Krakauer 2013).

This recasting of the evolutionary process as an inferential one[1] (Bergstrom and Rosvall 2009; Krakauer 2011) is based on the premise that organisms and other biological systems can be viewed as hypotheses about the present and future environments they or their offspring will encounter, induced from the history of past environmental states they or their ancestors have experienced. This premise, of course, only holds if the past is prologue—that is, has regularities, and the regularities can be estimated and even manipulated (as in niche construction) by biological systems or their components to produce adaptive behavior.

If these premises are correct, life at its core is computational, and a central question becomes: How do systems and their components estimate and control the regularity in their environments and use these estimates to tune their strategies? I suggest that the answer to this question, and the explanation for complexity, is that biological systems manipulate spatial and temporal structure to produce order—low variance—at local scales.

---

[1] This idea is related to work on Maxwell's demon (e.g., Krakauer 2011; Mandal, Quan, and Jarzynski 2013) and the Carnot cycle (e.g., Smith 2003), but we do not yet understand the mapping.

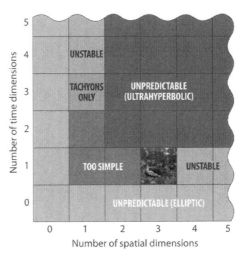

FIGURE 1. The dimensionality of the time–space continuum, with properties postulated when x does not equal 3 and y is larger than 1. Life on earth exists in three spatial dimensions and one temporal dimension. Biological systems effectively "discretize" time and space to reduce environmental uncertainty by coarse-graining and compressing environmental time series to find regularities. Components use the coarse-grained descriptions to predict the future, tuning their behavior to their predictions. [MODIFIED BY JAN UNDERWOOD FOR SFI FROM ORIGINAL BY MAX TEGMARK, WIKIMEDIA COMMONS]

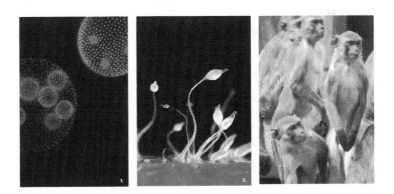

FIGURE 2. Biological systems—from (left to right) Volvox colonies, to slime molds, to animal societies, to large-scale ecosystems such as reefs, to human cities—are hierarchically organized, with multiple functionally important time and space scales. All feature: components with only partially aligned interests exhibiting coherent behavior at the aggregate level; components that turn over and that co-exist in the system at varying stages of development; social structure that persists but component behavior that fluctuates; and macroscopic variation in temporal and spatial structure and coupling with microscopic behavior, which has functional implications when the components can perceive—in evolutionary, developmental, or ecological time—regularities at the macroscopic scale.

PHASE II: UNIFIERS

Uncertainty Reduction

The story I want to tell starts with the observation that with each new level of organization typically comes new functionality—a new feature with positive payoff consequences for the system as a whole, or for its components (Flack, Erwin, Elliot, and Krakauer 2013). Policing in a pigtailed macaque group is an example. Once a heavy tailed distribution of power—defined as the degree of consensus in the group that an individual can win fights (see Flack and Krakauer 2006; Boehm and Flack 2010; Brush, Krakauer, and Flack 2013)—becomes effectively institutionalized (meaning hard to change) policing (an intrinsically costly strategy) becomes affordable, at least to those animals that sit in the tail of the power distribution: those superpowerful monkeys who are rarely or never challenged when they break up fights (Flack, de Waal, and Krakauer 2005; Flack, Girvan, de Waal, and Krakauer 2006).

My collaborators and I propose that a primary driver of the emergence of new functionality such as policing is the reduction of environmental uncertainty through the construction of nested dynamical processes with a range of characteristic time constants (Flack, Erwin, Elliot, and Krakauer 2013). These nested dynamical processes arise as components extract regularities from fast, microscopic behavior by coarse-graining (or compressing) the history of events to which they have been exposed.

Proteins, for example, can have a long half-life relative to RNA transcripts and can be thought of as the summed output of translation. Cells have a long half-life relative to proteins, and are a function of the summed output of arrays of spatially structured proteins. Both proteins and cells represent some average measure of the noisier activity of their constituents. Similarly, a pigtailed macaque's estimate of its power is a kind of average measure of the collective perception in the group that the macaque is capable of winning fights, and this is a better predictor of the cost the macaque will pay during fights than the outcome of any single melee, as these outcomes can fluctuate for contextual reasons. These coarse-grainings, or averages, are encoded as slow variables (Flack and de Waal 2007; Flack 2012; Flack, Erwin, Elliot, and Krakauer 2013; see also Feret, Danos, Krivine, Harner, and Fontana 2009, for a similar idea). Slow variables may have a spatial component as well as a temporal component, as in the protein and cell examples (fig. 6), or, minimally, only a temporal component, as in the monkey example.

As a consequence of integrating overabundant microscopic processes, slow variables provide better predictors of the local future configuration of a system than the states of the fluctuating microscopic components. In doing so, they promote accelerated rates of microscopic adaptation. Slow variables facilitate adaptation in two ways: they allow components to fine-tune their behavior, and they free components to search, at low cost, a larger space of strategies for extracting resources from the environment (Flack 2012; Flack, Erwin, Elliot, and Krakauer 2013). This phenomenon is illustrated by the power-in-support-of-policing example and also by work on the role of neutral networks in RNA folding. In the RNA case, many different sequences can fold into the same secondary structure. This implies that over evolutionary time, structure changes more slowly than sequence, thereby permitting sequences to explore many configurations under normalizing selection.

New Levels of Organization

As an interaction history builds up at the microscopic level, the coarse-grained representations of the microscopic behavior consolidate, becoming for the components increasingly robust predictors of the system's future state.

We speak of a new organizational level when the system's components rely to a greater extent on these coarse-grained or compressed descriptions of the system's dynamics for adaptive decision-making than on local fluctuations in the microscopic behavior

ENDOMESODERM SPECIFICATION UP TO 30 HOURS

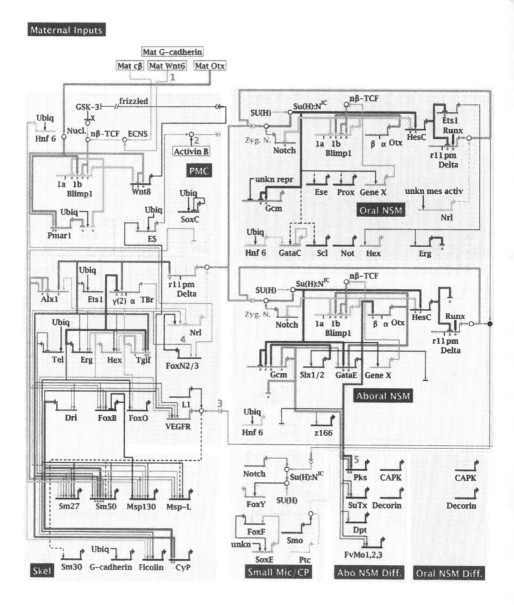

FIGURE 3. A sea urchin gene regulatory circuit. The empirically derived circuit describes the Boolean rules for coordinating genes and proteins to produce aspects of the sea urchin's phenotype—in this case, the position of cells in the endomesoderm at 30 hours since fertilization. Edges indicate whether a node induces a state change in another node, here genes and proteins. The circuit is a rigorous starting point for addressing questions about the logic of development and its evolution. In computational terms, the input is the set of relevant genes and proteins and the output is the target phenotypic feature.

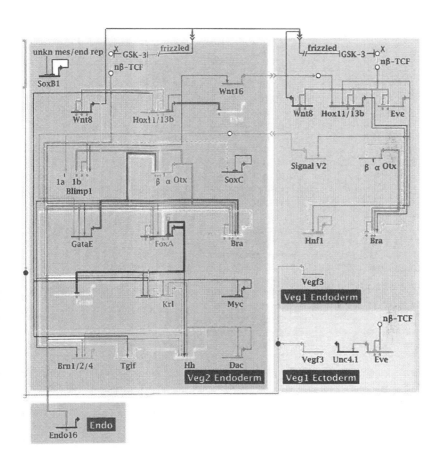

Ubiq = ubiquitous; Mat = maternal; activ = activator; rep = repressor; unkn = unknown; Nucl. = nuclearization; X = ß-catenin source;

nß-TCF = nuclearized b-ß-catenin-Tcf1; ES = early signal; ECNS = early cytoplasmic nuclearization system; Zyg. N. = zygotic Notch

Additional data sources for selected notes: 1: McClay lab; 2: Angerer lab; 3, 4: McClay lab; 5: Rogers and Calestani, 2010; 6: Croce and McClay

This model is frequently revised. It is based on the latest laboratory data [as of April 2014], some of which was not then published.

PHASE II: UNIFIERS

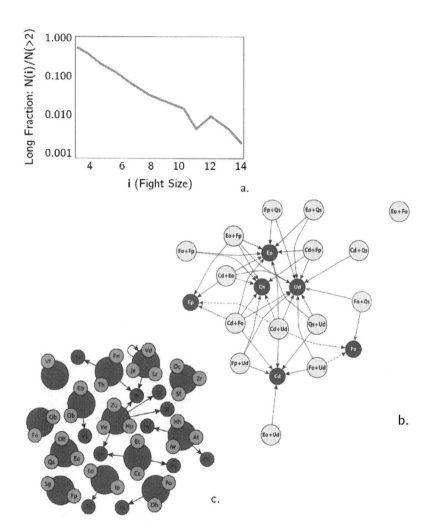

FIGURE 4. Cognitive effective theories for one macroscopic property of a macaque society: the distribution of fight sizes (a). To reduce circuit complexity we return to the raw time series data and remove as much noise as possible by compressing the data. In the case of our macaque dataset, this reveals which individuals and subgroups are regular and predictable conflict participants. We then search for possible strategies in response to these regular and predictable individuals and groups. This approach returns a family of circuits (b is an example), each of which has fewer nodes and edges than the full circuit (c). These circuits are simpler and more cognitively parsimonious. We then test the reduced circuits against each other in simulation to determine how well they recover the target macroscopic properties.

and when the coarse-grained estimates made by components are largely in agreement. The idea is that convergence on these "good enough" estimates underlies nonspurious correlated behavior among the components. This in turn leads to an increase in local predictability (e.g., Flack and de Waal 2007; Brush, Krakauer, and Flack 2013) and drives the construction of the information hierarchy. (Note that increased predictability can seem the product of downward causation in the absence of careful analysis of the bottom-up mechanisms that actually produced it.)

The Statistical Mechanics & Thermodynamics of Biology

Another way of thinking about slow variables is as a functionally important subset of the system's potentially many macroscopic properties. An advantage of this recasting is that it builds a bridge to physics, which over the course of its maturation as a field grappled with precisely the challenge now before biology: understanding the relationship between behavior at the individual or component level and behavior at the aggregate level.

In physics

As discussed in Krakauer and Flack (2010), the debate in physics began with thermodynamics—an equilibrium theory treating aggregate variables—and came to a close with the maturation of statistical mechanics—a dynamical theory treating microscopic variables.

Thermodynamics is the study of the macroscopic behavior of systems exchanging work and heat with connected systems or their environment. The four laws of thermodynamics all operate on average quantities defined at equilibrium—temperature, pressure, entropy, volume, and energy. These macroscopic variables exist in fundamental relationships with each other, as expressed, for example, in the ideal gas law. Thermodynamics is an extremely powerful framework as it provides experimentalists with explicit, principled recommendations

PHASE II: UNIFIERS

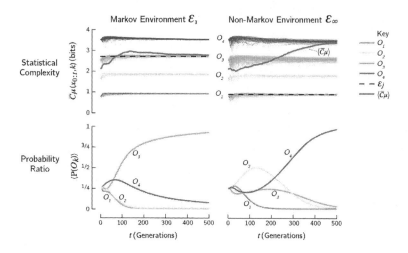

FIGURE 5. A comparison of Markov organisms in two environments: a Markov environment (left) and a non-Markov environment (right). In the top two plots, organismal complexity is plotted against time for each organism (organisms are represented by varying colors) and for many different sequences of 500 environmental observations; the bold red line shows the average organismal complexity, which in the Markov environment tends toward the environmental complexity and in the non-Markov environment exceeds it. In the bottom plots, the probability that a random organism has order k is plotted against time.

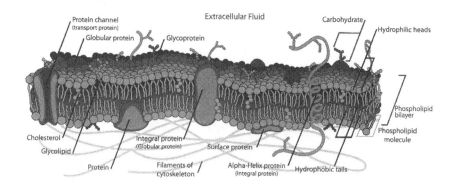

FIGURE 6. The cell can be thought of as a slow variable to the extent it is a function of the summed output of arrays of spatially structured proteins and has a long half-life compared to its proteins. Features that serve as slow variables provide better predictors of the local future configuration of a system than the states of the fluctuating microscopic components. We propose that when detectable by the system or its components, slow variables can reduce environmental uncertainty and, by increasing predictability, promote accelerated rates of microscopic adaptation.

about what variables should be measured and how they are expected to change relative to each other, but it is not a dynamical theory and offers no explanation for the mechanistic origins of the macroscopic variables it privileges. This is the job of statistical mechanics. By providing the microscopic basis for the macroscopic variables in thermodynamics, statistical mechanics establishes the conditions under which the equilibrium relations are no longer valid or expected to apply. The essential intellectual technologies behind much of statistical mechanics are powerful tools for counting possible microscopic configurations of a system and connecting these to macroscopic averages.

In biology

This brief summary of the relation between thermodynamics and statistical mechanics in physics is illuminating for two reasons. On the one hand it raises the possibility of a potentially deep division between physical and biological systems: so far—and admittedly biology is young—biology has had only limited success in empirically identifying important macroscopic properties and deriving these from first principles rooted in physical laws or deep evolved constraints.[2] This may be the case because many of the more interesting macroscopic properties are slow variables that result from the collective behavior of adaptive components, and their functional value comes from how components use them, making them fundamentally subjective (see Gell-Mann and Lloyd 1996 for more on subjectivity) and perhaps even nonstationary.[3]

On the other hand, the role of statistical mechanics in physics

---

[2] The work on scaling in biological systems shows a fundamental relationship between mass and metabolic rate, and this relationship can be derived from the biophysics (e.g., West, Brown, and Enquist 1997). Bettencourt and West are now investigating whether similar fundamental relationships can be established for macroscopic properties of human social systems, like cities (e.g., Bettencourt, Lobo, Helbing, Kuhnert, and West 2007; Bettencourt 2013).

[3] With the important caveat that in biology the utility of a macroscopic property as a predictor will likely increase as consensus among the components about the estimate increases, effectively reducing the subjectivity and increasing stationarity (see also Gell-Mann and Lloyd 1996).

PHASE II: UNIFIERS

suggests a way forward. If we have intuition about which macroscopic properties are important—that is, which macroscopic properties are slow variables—and we can get good data on the relevant microscopic behavior, we can proceed by working upward from dynamical many-body formalisms to equilibrium descriptions with a few favored macroscopic degrees of freedom (Levin, Grenfell, Hastings, and Perelson 1997; Krakauer and Flack 2010; Krakauer et al. 2011; Gintis, Doebeli, and Flack 2012).

A Statistical Mechanics–Computer Science–Information Theoretic Hybrid Approach

The most common approach to studying the relationship between micro and macro in biological systems is perhaps dynamical systems and, more specifically, pattern formation (for examples, see Sumpter 2006; Ball 2009; Couzin 2009; Payne et al. 2013). However, if, as we believe, the information hierarchy results from biological components collectively estimating regularities in their environments by coarse-graining or compressing time series data, a natural (and complementary) approach is to treat the micro and macro mapping explicitly as a computation.

> If we have intuition about which macroscopic properties are important—that is, which macroscopic properties are slow variables—and we can get good data on the relevant microscopic behavior, we can proceed by working upward from dynamical many-body formalisms to equilibrium descriptions with a few favored macroscopic degrees of freedom.

Elements of computation in biological systems

Describing a biological process as a computation minimally requires that we are able to specify the output, the input, and the algorithm or circuit connecting the input to the output (Flack and Krakauer 2011; see also Mitchell 2010; Valiant 2013). A secondary concern is how to determine when the desired output has been generated. In computer science this is called the termination criterion or halting problem. In biology it potentially can be achieved by constructing nested dynamical processes with a range of timescales, with the slower timescale processes providing the "background" against which a strategy is evaluated (Flack and Krakauer 2011), as discussed later in this paper in the section "Couplings."

A macroscopic property can be said to be an output of a computation if it can take on values that have functional consequences at the group or component level, if it is the result of a distributed and coordinated sequence of component interactions under the operation of a strategy set, and if it is a stable output of input values that converges (terminates) in biologically relevant time (Flack and Krakauer 2011). Examples studied in biology include aspects of vision such as edge detection (e.g., Olshausen and Field 2004), phenotypic traits such as the average position of cells in the developing endomesoderm of the sea urchin (e.g., Davidson 2010; Peter and Davidson 2011), switching in biomolecular signal-transduction cascades (e.g., Smith, Krishnamurthy, Fontana, and Krakauer 2011), chromatin regulation (e.g., Prohaska, Stadler, and Krakauer 2010), and social structures such as the distribution of fight sizes and the distribution of power in monkey societies (e.g., Flack 2012; Flack, Erwin, Elliot, and Krakauer 2013).

The input to the computation is the set of elements implementing the rules or strategies. As with the output, we do not typically know a priori which of many possible inputs is relevant, and so we must make an informed guess based on the properties of the output. In the case of the sea urchin's endomesoderm, we might start with a list of genes that have been implicated in the regulation

PHASE II: UNIFIERS

of cell position. In the case of the distribution of fight sizes in a monkey group, we might start with a list of individuals participating in fights.

Reconstructing the microscopic behavior

In biological systems the input plus the strategies constitute the system's microscopic behavior. There are many approaches to reconstructing the system's microscopic behavior. The most powerful is an experiment in which upstream inputs to a target component are clamped off and the output of the target component is held constant. This allows the experimentalist to measure the target component's specific contribution to the behavior of a downstream component (Pearl 2010). This type of approach is used to construct gene regulatory circuits mapping gene–gene and gene–protein interactions to phenotypic traits (fig. 3).

When such experiments are not possible, causal relationships can be established using time series analysis in which clamping is approximated statistically (Ay 2009; Pearl 2010). My collaborators and I have developed a novel computational technique, called Inductive Game Theory, that uses a variant of this statistical clamping principle to extract strategic decision-making rules, game structure, and (potentially) strategy cost from correlations observed in the time series data.

Collective computation through stochastic circuits

In all biological systems, of course, there are multiple components interacting and simultaneously coarse-graining to make predictions about the future. Hence the computation is inherently collective. A consequence of this is that it is not sufficient to simply extract from the time series the list of the strategies in play. We must also examine how different configurations of strategies affect the macroscopic output. One way these configurations can be captured is

Chapter 20: Life's Information Hierarchy

by constructing Boolean circuits describing activation rules as illustrated by the gene regulatory circuit shown in figure 3, which controls cell position (the output) at thirty hours from fertilization in the sea urchin (Peter and Davidson 2011). In the case of our work on micro to macro mappings in animal societies, we describe the space of microscopic configurations using stochastic "social" circuits (fig. 4).

Nodes in these circuits are the input to the computation. As discussed above, the input can be individuals or subgroups, or they can be defined in terms of component properties like age or neurophysiological state. A directed edge between two nodes indicates that the "receiving node" has a strategy for the "sending node"—and the edge weight can be interpreted as the above-null probability that the sending node plays the strategy in response to some behavior by the receiving node in a previous time step. Hence, an edge in these circuits quantifies the strength of a causal relationship between the behaviors of a sending and receiving node.

Sometimes components have multiple strategies in their repertoires. Which strategy is being played at time t may vary with context. These metastrategies can be captured in the circuit using different types of gates specifying how a component's myriad strategies combine (see also Feret, Davis, Krivine, Harmer, and Fontana 2009). By varying the types of gates and/or the strength of causal relationships, we end up with multiple alternative circuits—a family of circuits—all of which are consistent with the microscopic behavior, albeit with different degrees of precision. Each circuit in the family is essentially a model of the micro–macro relationship and so serves as a hypothesis for how strategies combine over nodes (inputs) to produce to the target output. We test the circuits against each other in simulation to determine which can best recover the actual measured macroscopic behavior of our system.

PHASE II: UNIFIERS

Cognitive effective theories for collective computation
The circuits describing the microscopic behavior can be complicated, with many "small" causes detailed, as illustrated by the gene regulatory circuit shown in figure 3. The challenge—once we have rigorous circuits—is to figure out the circuit logic (Flack and Krakauer 2011; see also Feret, Davis, Krivine, Harmer, and Fontana 2009).

There are many ways to approach this problem. Our approach is to build what's called in physics an effective theory: a compact description of the causes of a macroscopic property. Effective theories for adaptive systems composed of adaptive components require an additional criterion beyond compactness. As discussed earlier in this essay, components in these systems are tuning their behaviors based on their own effective theories—coarse-grained rules (see also Feret, Davis, Krivine, Harmer, and Fontana 2009)—that capture the regularities (Daniels, Krakauer, and Flack 2012). If we are to build an effective theory that explains the origins of functional space and timescales—new levels of organization—and ultimately the information hierarchy, the effective theory must be consistent with component models of macroscopic behavior, as these models, through their effects on strategy choice, drive that process. In other words, our effective theory should explain how the system itself is computing.

We begin the search for cognitively principled effective theories using what we know about component cognition to inform how we coarse-grain and compress the circuits. This means taking into account, given the available data, the kinds of computations components can perform and the error associated with these computations at the individual and collective levels, given component memory capacity and the quality of the "data sets" components use to estimate regularities (Krakauer, Flack, DeDeo, and Farmer 2010; Flack and Krakauer 2011; Daniels, Krakauer, and Flack 2012; all building on Gell-Mann 1996).

As we refine our understanding of the micro–macro mapping

through construction of cognitive effective theories, we also refine our understanding of what time series data constitute the "right" input—and hence the building blocks of our system. And, by investigating whether our best-performing empirically justified circuits can also account for other potentially important macroscopic properties, we can begin to establish which macroscopic properties might be fundamental and what their relation is to one another—the thermodynamics of biological collectives.

> The question we must answer is, What is the optimal coupling between macroscopic and microscopic change, and can systems, by manipulating how components are organized in space and time, get close to this optimal coupling?

Couplings, information flow, and macroscopic tuning
Throughout this essay I have stressed the importance of slowness (effective stationarity) for prediction. Slowness also has costs, however. Consider our power example. The power structure must change slowly if individuals are to make worthwhile investments in strategies that work well given the structure, but it cannot change too slowly or it may cease to reflect the underlying distribution of fighting abilities on which it is based and, hence, cease to be a good predictor of interaction cost (Flack 2012; Flack, Erwin, Elliot, and Krakauer 2013). The question we must answer is, What is the optimal coupling between macroscopic and microscopic change, and can systems, by manipulating how components are organized in space and time, get close to this optimal coupling?

PHASE II: UNIFIERS

One approach to this problem is to quantify the degeneracy of the target macroscopic property and then perturb the circuits by either removing nodes, up- or down-regulating node behavior, or restructuring higher-order relationships (subcircuits) to determine how many changes at the microscopic level need to occur to induce a state change at the macroscopic level.

Another approach is to ask how close the system is to a critical point—that is, how sensitive the target macroscopic property is to small changes in parameters describing the microscopic behavior. Many studies suggest that biological systems of all types sit near the critical point (Mora and Bialek 2011). A hypothesis we are exploring is that sitting near the critical point means that important changes at the microscopic scale will be visible at the macroscopic scale. Of course this also has disadvantages as it means small changes can potentially cause big institutional shifts, undermining the utility of coarse-graining and slow variables for prediction (Flack, Erwin, Elliot, and Krakauer 2013).

If balancing trade-offs between robustness and prediction on the one hand, and adaptability to changing environments on the other, can be achieved by modulating the coupling between scales (Flack, Hammerstein, and Krakauer 2012; Flack, Erwin, Elliot, and Krakauer 2013), we should be able to make predictions about whether a system is far from, near, or at the critical point based on whether the data suggest that robustness or adaptability is more important given the environment and its characteristic timescale. This presupposes that the system can optimize where it sits with respect to the critical point, implying active mechanisms for modulating the coupling. We are working to identify plausible mechanisms using a series of toy models to study how the type of feedback from the macroscopic or institutional level to the microscopic behavior influences the possibility of rapid institutional switches.

Complexity

This essay covers a lot of work, so allow me to summarize. I suggested that the origins of the information hierarchy lie in the manipulation of space and time to reduce environmental uncertainty. I further suggested that uncertainty reduction is maximized if the coarse-grained representations of the data the components compute are in agreement (because this increases the probability that everyone is making the same predictions and so tuning the same way). As this happens, the coarse-grained representations consolidate into robust, slow variables at the aggregate level, creating new levels of organization and giving the appearance of downward causation.

I proposed that a central challenge lies in understanding what the mapping is between the microscopic behavior and these new levels of organization. (How exactly do everyone's coarse-grainings converge?) I argued that, in biology, a hybrid statistical mechanics–computer science–information theoretic approach (see also Krakauer et al. 2011) is required to establish such mappings. Once we have cognitively principled effective theories for mappings, we will have an understanding of how biological systems, by discretizing space and time, produce information hierarchies.

Where are we, though, with respect to explaining the origins of biological complexity?

The answer we are moving toward lies at the intersection of the central concepts in this essay. If evolution is an inferential process with complex life being the result of biological systems extracting regularities from their environments to reduce uncertainty, a natural recasting of evolutionary dynamics is in Bayesian terms. Under this view, organism and environment can be interpreted as k-order Markov processes and modeled using finite-state hidden Markov models (fig. 5). Organisms update prior models of the environment with posterior models of observed regularities. We are exploring how the Markov order (a proxy for memory) of organisms changes as organisms evolve to match their environment, quantifying fit

PHASE II: UNIFIERS

to the environment with model selection. We use information-theoretic measures to quantify structure. Our approach allows us to evaluate the memory requirements of adapting to the environment given its Markov order, quantify the complexity of the models organisms build to represent their environments, and quantitatively compare organismal and environmental complexity as our Markov organisms evolve. We hypothesize that high degrees of complexity result when there is regularity in the environment, but it takes a long history to perceive it and an elaborate model to encode it. ❦

ACKNOWLEDGMENTS

This essay summarizes my view of the past, present, and predicted future of the core research program at the Center for Complexity & Collective Computation (C4). In addition to our current collaborators—Nihat Ay, Dani Bassett, Karen Page, Chris Boehm, and Mike Gazzaniga—and the super-smart students and postdoctoral fellows Eleanor Brush, Bryan Daniels, Simon DeDeo, Karl Doron, Chris Ellison, the late Tanya Elliot, Evandro Ferrada, Eddie Lee, and Philip Poon, who have carried out much of this work on a daily basis, I am deeply grateful to the Santa Fe Institute for its support over the years and to the Santa Fe folks whose ideas have provided inspiration. First and foremost this includes David Krakauer, my main collaborator. Other significant SFI influences include Jim Crutchfield, Doug Erwin, Walter Fontana, Lauren Ancel Meyers, Geoffrey West, Eric Smith, Murray Gell-Mann, Bill Miller, David Padwa, and Cormac McCarthy. I am indebted to Ellen Goldberg for making possible my first postdoctoral position at SFI. Finally, much of this research would not be possible without the generous financial support provided by the John Templeton Foundation through a grant to SFI to study complexity and a grant to C4 to study the mind–brain problem, a National Science Foundation grant (0904863), and a grant from the US Army Research Laboratory and the US Army Research Office under contract number W911NF-13-1-0340.

Chapter 20: Life's Information Hierarchy

REFERENCES CITED

Ancel, L. W., and W. Fontana. 2000. "Plasticity, Evolvability, and Modularity in RNA." *Journal of Experimental Zoology. Part B, Molecular and Developmental Evolution* 288: 242–83.

Ay, N. 2009. "A Refinement of the Common Cause Principle." *Discrete Applied Mathematics* 157: 2439–57.

Ball, P. 2009. *Nature's Patterns: A Tapestry in Three Parts*. Oxford, UK: Oxford University Press.

Bergstrom, C. T., and M. Rosvall. 2009. "The Transmission Sense of Information." *Biology and Philosophy* 26: 159–76.

Bettencourt, L. M. A. 2013. "The Origins of Scaling in Cities." *Science* 340: 1438–41.

Bettencourt, L. M. A., J. Lobo, D. Helbing, C. Kuhnert, and G. B. West. 2007. "Growth, Innovation, Scaling, and the Pace of Life in Cities." *PNAS* 104: 7301–6.

Boehm, C., and J. C. Flack. 2010. "The Emergence of Simple and Complex Power Structures through Niche Construction." In *The Social Psychology of Power*, edited by A. Guinote and T. K. Vescio, 46–86. New York, NY: Guilford Press.

Brush, E. R., D. C. Krakauer, and J. C. Flack. 2013. "A Family of Algorithms for Computing Consensus About Node State from Network Data." *PLOS Computational Biology* 9: e1003109.

Buss, L. W. 1987. *The Evolution of Individuality*. Princeton, NJ: Princeton University Press.

Couzin, I. D. 2009. "Collective Cognition in Animal Groups." *Trends in Cognitive Science* 13: 36–43.

Crutchfield, J. P. 1994. "The Calculi of Emergence: Computation, Dynamics, and Induction." *Physica D* 75: 11–54.

Crutchfield, J. P., and D. P. Feldman. 2001. "Synchronizing to the Environment: Information-Theoretic Constraints on Agent Learning." *Advances in Complex Systems* 4: 251–64.

Crutchfield, J. P., and O. Görnerup. 2006. "Objects That Make Objects: The Population Dynamics of Structural Complexity." *Journal of the Royal Society Interface* 22: 345–49.

Daniels, B., D. C. Krakauer, and J. C. Flack. 2012. "Sparse Code of Conflict in a Primate Society." *PNAS* 109: 14259–64.

Daniels, B., D. C. Krakauer, and J. C. Flack. N.d. *Conflict Tuned to Maximum Information Flow*. In preparation.

PHASE II: UNIFIERS

Davidson, E. H. 2010. "Emerging Properties of Animal Gene Regulatory Networks." *Nature* 468: 911–20.

Dedeo, S., D. C. Krakauer, and J. C. Flack. 2010. "Inductive Game Theory and the Dynamics of Animal Conflict." *PLOS Computational Biology* 6: E1000782.

Ellison, C., J. C. Flack, and D. C. Krakauer. n.d. *On Inferential Evolution and the Complexity of Life*. In preparation.

Feldman, M., and I. Eschel. 1982. "On the Theory of Parent–Offspring Conflict: A Two-Locus Genetic Model." *American Naturalist* 119: 285–92.

Feret, J., V. Danos, J. Krivine, R. Harmer, and W. Fontana. 2009. "Internal Coarse-Graining of Molecular Systems." *PNAS* 106: 6453–58.

Ferrada, E., and D. C. Krakauer. n.d. *The Simon Modularity Principle*. In preparation.

Flack, J. C. 2012. "Multiple Time-Scales and the Developmental Dynamics of Social Systems." *Philosophical Transactions of the Royal Society B: Biological Sciences* 367: 1802–10.

Flack, J. C., and D. C. Krakauer. 2006. "Encoding Power in Communication Networks." *American Naturalist* 168: E87–102.

Flack, J. C., and F. B. M. De Waal. 2007. "Context Modulates Signal Meaning in Primate Communication." *PNAS* 104: 1581–86.

Flack, J. C., and D. C. Krakauer. 2011. "Challenges for Complexity Measures: A Perspective from Social Dynamics and Collective Social Computation." *Chaos* 21: 037108.

Flack, J. C., F. B. M. De Waal, and D. C. Krakauer. 2005. "Social Structure, Robustness, and Policing Cost in a Cognitively Sophisticated Species." *American Naturalist* 165: E126–39.

Flack, J. C., P. Hammerstein, and D. C. Krakauer. 2012. "Robustness in Biological and Social Systems." In *Evolution and the Mechanisms of Decision-Making*, edited by P. Hammerstein, and J. Stevens, 129–51. Cambridge, MA: MIT Press.

Flack, J. C., D. Erwin, T. Elliot, and D. C. Krakauer. 2013. "Timescales, Symmetry, and Uncertainty Reduction in the Origins of Hierarchy in Biological Systems." In *Cooperation and Its Evolution*, edited by K. Sterelny, R. Joyce, B. Calcott, and B. Fraser, 45–74. Cambridge, MA: MIT Press.

Flack, J. C., M. Girvan, F. B. M. De Waal, and D. C. Krakauer. 2006. "Policing Stabilizes Construction of Social Niches in Primates." *Nature* 439: 426–29.

Fontana, W., and L. W. Buss. 1996. "The Barrier of Objects: From Dynamical Systems to Bounded Organizations." In *Boundaries and Barriers*, edited by J. Casti and A. Karlqvist, 56–116. Reading, MA: Addison–Wesley.

Fontana, W., and P. Schuster. 1998. "Continuity in Evolution: On the Nature of Transitions." *Science* 280: 1451–55.

Frank, S. A. 2003. "Repression of Competition and the Evolution of Cooperation." *Evolution* 57: 693–705.

Gell-Mann, M. 1996. "Fundamental Sources of Unpredictability." Talk presented at conference of the same name. http://www-physics.mps.ohio-state.edu/~perry/p633_sp07/articles/fundamental-sources-of-unpredictability.pdf. Accessed 01/10/2014.

Gell-Mann, M., and S. Lloyd. 1996. "Information Measures, Effective Complexity, and Total Information." *Complexity* 2: 44–53.

Gintis, H., M. Doebeli, and J. C. Flack. 2012. "The Evolution of Human Cooperation." *Cliodynamics: Journal of Theoretical and Mathematical History* 3.

Krakauer, D. C. 2011. "Darwinian Demons, Evolutionary Complexity, and Information Maximization." *Chaos* 21: 037110.

Krakauer, D. C., N. Bertschinger, N. Ay, E. Olbrich, and J. C. Flack. n.d. "The Information Theory of Individuality." In *What Is an Individual?*, edited by L. Nyhart and S. Lidgard. Chicago, IL: University of Chicago Press. In review.

Krakauer, D. C., and J. C. Flack. 2010. "Better Living Through Physics." *Nature* 467: 661.

Krakauer, D. C., and P. Zanotto. 2009. "Viral Individuality and Limitations of the Life Concept." In *Protocells: Bridging Non-Living and Living Matter*, edited by M. A. Rasmussen et al., 513–36. Cambridge, MA: MIT Press.

Krakauer, D. C., J. C. Flack, S. DeDeo, and D. Farmer. 2010. "Intelligent Data Analysis of Intelligent Systems." *IDA 2010 LNCS* 6065: 8–17.

Krakauer, D. C., J. P. Collins, D. Erwin, J. C. Flack, W. Fontana, M. D. Laubichler, S. Prohaska, G. B. West, and P. Stadler. 2011. "The Challenges and Scope of Theoretical Biology." *Journal of Theoretical Biology* 276: 269–76.

Lee, E., B. Daniels, D. C. Krakauer, and J. C. Flack. n.d. *Cognitive Effective Theories for Probabilistic Social Circuits Mapping Strategy to Social Structure*. In preparation.

PHASE II: UNIFIERS

Levin, S. A., B. Grenfell, A. Hastings, and A. S. Perelson. 1997. "Mathematical and Computational Challenges in Population Biology and Ecosystems Science." *Science* 275: 334–43.

Mandal, D., H. T. Quan, and C. Jarzynski. 2013. "Maxwell's Refrigerator: An Exactly Solvable Model." *Physical Review Letters* 111: 030602.

Michod, R. E. 2000. *Darwinian Dynamics: Evolutionary Transitions in Fitness and Individuality.* Princeton, NJ: Princeton University Press.

Mitchell, M. 2010. "Biological Computation." Working Paper 2010-09-021, Santa Fe Institute, Santa Fe, NM.

Mora, T., and W. Bialek. 2011. "Are Biological Systems Poised at Criticality?" *Journal of Statistical Physics* 144: 268–302.

Olshausen, B., and D. Field. 2004. "Sparse Coding of Sensory Inputs." *Current Opinion in Neurobiology* 14: 481–87.

Payne, S., L. Bochong, Y. Cao, D. Schaeffer, M. D. Ryser, and L. You. 2013. "Temporal Control of Self-Organized Pattern Formation Without Morphogen Gradients in Bacteria." *Molecular Systems Biology* 9: 697.

Pearl, J. 2010. *Causality,* 2nd ed. Cambridge, MA: Cambridge University Press.

Peter, I. S., and E. H. Davidson. 2011. "A Gene Regulatory Network Controlling the Embryonic Specification of Endoderm." *Nature* 474: 635–39.

Poon, P., J. C. Flack, and D. C. Krakauer. n.d. *Niche Construction and Institutional Switching Via Adaptive Learning Rule*s. In preparation.

Prohaska, S. J., P. F. Stadler, and D. C. Krakauer. 2010. "Innovation in Gene Regulation: The Case of Chromatin Computation." *Journal of Theoretical Biology* 265: 27–44.

Sabloff, J. A., ed. n.d. *The Rise of Archaic States: New Perspectives on the Development of Complex Societies.* Santa Fe Institute. In preparation.

Schuster, P., and W. Fontana. 1999. "Chance and Necessity in Evolution: Lessons from RNA." *Physica D: Nonlinear Phenomena* 133: 427–52.

Smith, E. 2003. "Self-Organization from Structural Refrigeration." *Physical Review E* 68: 046114.

———. 2008. "Thermodynamics of Natural Selection I: Energy Flow and the Limits on Organization." *Journal of Theoretical Biology,* http://dx.doi.org/10.1016/j.jtbi.2008.02.010, PDF.

Smith, E., S. Krishnamurthy, W. Fontana, and D. C. Krakauer. 2011. "Nonequilibrium Phase Transitions in Biomolecular Signal Transduction." *Physical Review E* 84: 051917.

Smith, J. M., and E. Szathmáry. 1998. *The Major Transitions in Evolution.* Oxford: Oxford University Press.

Stadler, B. M. R., P. F. Stadler, G. Wagner, and W. Fontana. 2001. "The Topology of the Possible: Formal Spaces Underlying Patterns of Evolutionary Change." *Journal of Theoretical Biology* 213: 241–74.

Sumpter, D. J. T. 2006. "The Principles of Collective Animal Behaviour." *Philosophical Transactions of the Royal Society B* 361: 5–22.

Valentine, J., and R. May. 1996. "Hierarchies in Biology and Paleontology." *Paleobiology* 22: 23–33.

Valiant, L. 2013. *Probably Approximately Correct.* New York, NY: Basic Books.

West, G. B., J. H. Brown, and B. J. Enquist. 1997. "A General Model for the Origin of Allometric Scaling Laws in Biology." *Science* 276: 122–26.

2015 AND BEYOND

TERRAFORMERS

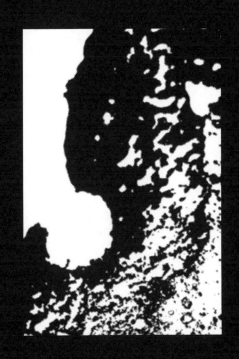

COMPLEXITY: WORLDS HIDDEN IN PLAIN SIGHT

David C. Krakauer, SFI
Christian Science Monitor, November 19, 2015

On August 14, 2003, fifty million people across the American Northeast lost electrical power. Around 70,000 megawatts went AWOL, which for reference is on the order of one hundred coal plants spontaneously disappearing from energy production.

The initial cause was thought to be a fault in a 345-kilovolt line in northern Ohio. In less than an hour, this fault led to further lines being lost due to excessive load propagating through the power grid. This negative-electrical epidemic spread across a considerable part of the nation and led to a regional blackout.

Further investigation pointed to additional "causes." A parallel and compounding problem was a software failure in a control center. Another was that trees that had been supporting electrical cables had grown too tall, promoting short circuits. Another was that the utility companies managing parts of the grid did not have as effective or efficient a communication strategy as is needed under pressure.

The blackout inflicted an economic cost of around $10 billion.

To summarize, a software network interacted with a physical cable network supported by a forest ecological network overseen by a stressed human social network. The failure was not "disciplinary" or "departmental;" it was complex. A full understanding of one critical infrastructure, the power grid, requires an understanding of a multitude of overlapping networks.

Complexity science is an effort to discern and theorize common patterns in complex systems from multiple scientific perspectives. Many scientific disciplines are already associated with powerful models and theories: in biology, for example, there is the theory

PHASE III: TERRAFORMERS

of evolution; in economics there is utility maximization and game theory; and in engineering mathematics there is Alan Turing's theory of computation.

Complexity science seeks to connect these theories, to find explanatory and predictive frameworks that allow us to, for example, describe biological mechanisms in computational terms or social structures in energetic terms.

For the last few decades we have been steadily surveying the landscape of complex phenomena, and it is gratifying that along the way we find that complex systems nominally unrelated bear strong family resemblances. These similarities include how the mathematical structure of evolutionary adaptation looks a lot like the mathematics of learning, that the distribution of energy within a body made of tissues and fluids follows rules similar to those governing the distribution of energy in a society, that networks within cells adhere to the geometric principles we find on the internet, and that the rise and fall of ancient civilizations follow a sequence similar to the extraordinary growth and contraction of urban centers we see in our own millennium.

It comes as something of a surprise, though, that many of the systems we scientists understand the best are those that we shall never touch (the sun), never see (the quark), and never feel (the Higgs field). My Santa Fe Institute colleague and Nobel laureate in physics Murray Gell-Mann captured the essence of this achievement in his Nobel Prize banquet speech in 1969: "How can it be that writing down a few simple and elegant formulae, like short poems governed by strict rules such as those of the sonnet or the Waka, can predict universal regularities of Nature?"

Compare this to the world in which we live, the biological, psychological, social, and cultural domains with which we enjoy direct sensory experience. This complex world continues to elude the compressive eloquence of the formulae of physics and chemistry,

Chapter 21: Complexity: Worlds Hidden in Plain Sight

with their uncanny resemblance to the elegant artistic properties of the Japanese verse form, the Waka.

What is it that makes observable complex systems such as the economy, sustainable urbanization, or human conflict so challenging?

This paradox of comprehension was articulated explicitly by a great physicist of an earlier age: "Sir Isaac Newton, when asked what he thought of the infatuations of the people, answered that he could calculate the motions of erratic bodies, but not the madness of a multitude" (quoted from *The Church of England Quarterly Review*, 1850).

It seems that, from the perspective of mathematical science, there exist two natural domains. The first is the physical domain of particles, fields, and universal laws, with an associated search for elegant theories that apply everywhere in the known universe. Here, science has made great strides.

It comes as something of a surprise, though, that many of the systems we scientists understand the best are those that we shall never touch (the sun), never see (the quark), and never feel (the Higgs field).

The second domain is that of complex phenomena. These are adaptive, interacting, many-body systems that include populations of cells, societies, economies, cities, human cultures, and technological networks—all phenomena with long histories and adaptive components, and they have a tendency to change as soon as we have come to understand them. Complexity theories extend to life—a

PHASE III: TERRAFORMERS

remarkable state hitherto found only upon the crust of our third planet from the sun.

As with physical theory (such as the theory of gravity, which we need to understand if we are to make any progress with ballistics, aviation, and space flight), some form of complexity theory is required if we are to understand many of the intimate, and patently uncertain, interactions found in modern society. And the natural complement to the search for fundamental theory is the direct and ancillary discovery of tools to predict and control the complex, highly interconnected world in which we live.

Many of our most pressing challenges and failures in the twenty-first century derive from an underestimation of complexity. Society has a tendency to treat challenges as if they emerged from a single factor in a rather straightforward way. Hence, we blame war on a single aggressor, starvation on the scarcity of a single food product, or poverty on the concentration of wealth.

The falsehood that intellectual scholarship is incompatible with practical reality is one of the most pernicious modern myths propagated by the dataphobic and the political ideologues. We shall challenge this perspective.

The temptation to avoid complexity is rather firmly rooted in the abiotic, physical domain, where certainty reigns. And if this were not bad enough, our educational system tends to perpetuate

Chapter 21: Complexity: Worlds Hidden in Plain Sight

this misunderstanding with departments and schools that treat the interconnected world around us as if it was simple and disconnected.

In this series of invited articles for the *Christian Science Monitor* [the essays that follow in this volume], my colleagues at the Santa Fe Institute shall review the recent progress in complexity science. We shall examine the most fundamental, brain-tickling mathematical and computational theories and methods coming out of this endeavor.

The falsehood that intellectual scholarship is incompatible with practical reality is one of the most pernicious modern myths propagated by the dataphobic and the political ideologues. We shall challenge this perspective.

Complexity science provides beautiful examples of hard technical problems engaging in a productive exchange with some of our thorniest societal dilemmas. The future prosperity of life on Earth, and, at some point in the human trajectory, beyond the earth, require that we engage to the maximum extent possible with our precious faculty of reason to better understand the workings of complexity and, one day, Newton's "madness of the multitudes."

PHASE III: TERRAFORMERS

All complex adaptive systems involve learning and adaptation by individuals and collectives, and, as a result, these systems develop intricate and unique histories. This is why Mumbai is so different from Seattle. Each is the result of its particular geography, its culture, its mixture of peoples, its politics, its accidents of history.

From this, you might assume that cities are all idiosyncratic agglomerations of humanity, and that the common essence of "cityness" can never be found. Not so. When we quantitatively compare the properties of Mumbai and Seattle, or New York City and Santa Fe, we find that all cities, from the tiniest hamlets to the largest megalopolises, share certain general spatial and socioeconomic features.

> **All complex adaptive systems involve learning and adaptation by individuals and collectives, and, as a result, these systems develop intricate and unique histories.**

To study this underlying order, we and other researchers analyze lots of data to compare many different properties of thousands of cities around the world. Analysis on this vast scale is necessary to transcend the specificities of each place, though only now is it becoming possible to do this systematically as data on human societies become easier to collect and share.

We analyze just about anything that can be measured: A city's total area, the extent of its roads and highways, its lights at night, its gross domestic product (GDP), the wages and professions of its workers, its crime rate, and much more.

Then, we organize all these diverse quantities, typically studied separately by researchers in different academic disciplines, under a unifying complex systems framework. We do this by thinking of social and economic activity as a gigantic network of people and their interactions embedded into the physical built spaces of cities, which comprise places (buildings, public areas) and access networks (streets, pipes, electrical and telephone cables, internet hubs).

In these terms, all cities share a general shape (a "topology"), despite individual differences in the geometry of streets and city blocks. The general characteristics of these spaces allow us to estimate how often people meet, and thus the rate at which they can produce socioeconomic outputs ranging from ideas and wealth to crime and disease.

A Scaling Theory of Cities

And here's where it gets interesting: as cities grow and their networks evolve, the area or volume of the networks needed to keep them functionally connected tends to become smaller on a per capita basis. For example, in larger cities more people can share the same bus or segment of road or sewer pipe.

This introduces two important concepts: first, the idea of *scaling*, which refers to how measurable properties of a system change with its size; second, the concept of *economies of scale*. The latter means that, as cities grow, they need less of something per person: roads, sewers, or gas stations, for example. What's more, such economies of scale in infrastructure are, mathematically speaking, regular and predictable, given a city's size: if you double the size of a city, you only need 80–90 percent more street surfaces, gas stations, sewer lines, etc. This holds true, within a few percentage points, for cities of all sizes and across all nations.

This increase is nevertheless faster than the increase in the overall land area the city occupies. This is why roads, cables, and pipes become so ubiquitous in larger cities and eventually must be buried underground, especially in a city's densest parts. This is also why roads in

PHASE III: TERRAFORMERS

larger cities are more congested and often under construction: they are used much more intensely than in smaller towns.

The converse of economies of scale—*increasing returns to scale*—gets us closer to why cities exist. The size of a city's economy (measured as its GDP), for example, is typically larger per capita in a bigger city. This is why New York City is so expensive, but it is also why New Yorkers make more money on average than people living in a city half its size (Los Angeles) and by a predictable amount (10–20 percent more).

Increasing returns characterize most outputs of human social interaction in a city, manifesting itself in predictably higher wages, faster technological innovation, more congested streets, and more incidents of violent crime per person in larger urban areas.

As we become networked on ever-larger scales, we can exploit the division of knowledge and labor to create organizations that are more efficient and that, collectively, contain and process more knowledge. Cities are a general-purpose way to build such networks and to rev up the engines of growth and change in human societies.

The integration of all these urban quantities suggests that these two effects—economies of scale and increasing returns to scale—are not independent but rather mirror images of each other. In this way, the geometry of our social lives is, in a very formal sense, interdependent of the spaces we build in cities.

Cities as Social Reactors

Another, more intuitive way to understand urban scaling is the speed of life in cities. Life in larger cities is generally faster, meaning one can do more, both good and bad, by having more social exchanges over the same amount of time. This speed of social life is intimately connected to the spatial density of many activities, creating spaces and times when a lot can happen quickly.

This reveals the ultimate function of cities: cities are social reactors, places where interactions among many different strangers can be realized and sustained. Ultimately, it is this accelerating dynamic that creates the buzz of a great city.

As economists have known for a long time, this continuous feedback between people's productive activities and new ideas is the driver of economic growth. As we become networked on ever-larger scales, we can exploit the division of knowledge and labor to create organizations that are more efficient and that, collectively, contain and process more knowledge. Cities are a general-purpose way to build such networks and to rev up the engines of growth and change in human societies.

But here's the rub: this magical power of cities does not always work well or apply equally to everybody. All the quantities for which we find scaling relationships are merely averages—quantitative snapshots of entire metropolitan areas. Being able to predict the average per capita income in New York City (about $60,000 per year) is obviously important for understanding the dynamics of cities, but equally important is to understand why it has the same income inequality as Port-au-Prince, Haiti, which of course is much poorer.

The Challenge of Cities

It's clear that the increased social opportunity of larger cities affects different people in different ways, young and old, rich and poor. This results in very large inequalities and heterogeneities

PHASE III: TERRAFORMERS

across any large city, a phenomenon still poorly understood and often mismanaged.

In our own cities, and certainly in developing cities throughout the world, it is common for poverty and exclusion to coexist with wealth and opportunity. Even as cities continue to grow, many people in developing countries lack access to even the most basic services such as clean water, modern sanitation, or basic justice. The imperative of meeting basic necessities on a daily basis leaves little time for educational pursuits, and much less for sophisticated entrepreneurship.

Achieving these goals may well represent the only way to save ourselves and the planet from major environmental and humanitarian crises and thereby set the course for a world of open-ended creativity and urbanized human development.

This is the greatest obstacle to human development for some one billion people on our planet today, a number that could triple by 2050 if no practical policy solutions are devised to tackle these issues systematically.

To deal with these challenges, representatives from all 193 countries of the United Nations General Assembly signed a remarkable blueprint for the future in September that commits us to worldwide Sustainable Development Goals to be met over the next fifteen years. These goals set quantitative local targets to eliminate extreme poverty, provide universal access to basic services, institute higher standards of justice and government, and achieve urban growth that is environmentally sustainable.

Achieving these goals may well represent the only way to save ourselves and the planet from major environmental and humanitarian crises and thereby set the course for a world of open-ended creativity and urbanized human development.

The complexity and urgency of these challenges are colossal, far outweighing the Apollo Program or the Manhattan Project. Fifteen years is not a lot of time. But the key to their solution is to understand the systemic nature of these problems defined by the lives and economic activities of people, and by the environments and services that we build in cities. To continue to seize the transformative power of cities, and to do it fast, we need to match extraordinary action with holistic and exacting knowledge that captures the nature of cities as complex adaptive systems.

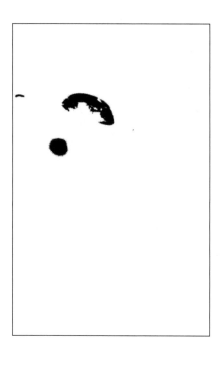

PREDICTING THE NEXT RECESSION

*Rob Axtell, George Mason University, and
J. Doyne Farmer, Institute for New Economic Thinking
Christian Science Monitor, December 10, 2015*

The 2008 financial crisis—which cost the United States economy between $6 trillion and $14 trillion and the world economy a great deal more—shook the world of finance to its foundation.

It hit the most vulnerable particularly hard, as unemployment in the United States doubled from 5 to 10 percent, and, in several countries in southern Europe, one in four people who want a job are still unable to find work—roughly the unemployment rate the US experienced during the Great Depression.

It's now old hat to point out that very few experts saw it coming. We shouldn't be too hard on them, though. Surprisingly, the US investment in developing a better theoretical understanding of the economy is very small—around $50 million in annual funding from the National Science Foundation—or just 0.0005 percent of a $10 trillion crisis.

With the Eurozone crisis still unfolding and financial panics now a regular occurrence on Wall Street, the next trillion-dollar meltdown might not be that far away. We justifiably spend billions trying to understand the weather, the climate, the oceans, and the polar regions. Why is the budget for basic research in economics, something that touches us all directly and daily, so paltry?

We think this has something to do with the way economics research is traditionally done, and we have a better way: loading millions of artificial households, firms, and people into a computer and watching what happens when they are allowed to interact.

PHASE III: TERRAFORMERS

Extreme Events

We've already made considerable progress with agent-based models in economics. They provide the best explanations, for example, for the volatility of prices in financial markets, explaining why there are periods in which markets heat up and others when they cool down.

They also provide the best model for the prevalence of extreme events: why crashes occur more frequently than one would expect. We've built agent-based models of firm dynamics and labor movements between firms that explain dozens of features of the US economy.

In other fields, such as traffic analysis and public health, agent-based modeling has become the standard approach when policy-relevant solutions are needed. Agent-based traffic models offer planners detailed representations of the causes of congestion in cities. In Portland, Oregon, for example, such modeling simulated traffic on every street in the city.

In economics, however, agent-based modeling is still the new kid on the block. We think it is time to give the complex systems approach a serious try. We need to build detailed agent-based models of economies.

Diversifying our Theory Portfolio

The type of model we envision can represent the behavior of the full variety of American households, from farmers to executives, retirees to recent college graduates. It can simulate the activities of a wide variety of firms, from large manufacturers to small businesses. It can include data on how the largest banks operate and, as a matter of course, alert regulators when the risks banks are taking rise to the systemic level, jeopardizing the operation of the whole financial system.

The goal of large-scale agent-based models is not to make point predictions: they do not say, for example, where the economy will be, exactly, in six months. Nor could they be used to trade on financial markets.

Rather, after a simulation is run a thousand or a million times, such models show policy makers the range of possible futures and their relative probabilities if policies remain unchanged. Then, by altering policies in the model and running them another million times, we might begin to understand paths to better futures.

Today we have outdated, twentieth-century models for managing an unpredictable twenty-first-century economy. Only through quantitative models will economic policy disagreements have a chance to be settled via the scientific method, turning philosophical arguments into discussions of model parameters.

In the meantime, we remain vulnerable to trillion-dollar meltdowns. A reasonable strategy to avoid bad economic outcomes in our collective future involves broad investment in a diverse portfolio of economic models in hopes our policy makers can get better guidance than they had the last time.

ARE HUMANS TRULY UNIQUE? HOW DO WE KNOW?

Jennifer A. Dunne and Marcus J. Hamilton, SFI
Christian Science Monitor, December 23, 2015

Of the millions of species on the planet, humans seem fairly unique. We have produced flamenco dancing and skinny jeans, jet skis and cell phones, New York and Tokyo, and fabulous art and music by creative individuals from Leonardo da Vinci to Joni Mitchell.

While many aspects of human technology, culture, and society have no clear counterparts in other species, do they make us truly unique? Every species possesses traits that other species don't, which is how we distinguish a ferret from a starfish. If every species is different, what, if anything, sets the human species uniquely apart from other species?

The Energetic Imperative

The science of complex systems, or complexity science, offers powerful tools for assessing human uniqueness. One way to compare all species, including humans, is energy consumption. Energy is the fundamental currency of life, as all species use energy to grow, survive, and reproduce.

On a biological level, people are no different from other species. Joni Mitchell's first energetic imperative is to take in a sufficient number of calories every day to maintain her metabolic function and to fuel her basic activity, and only with the excess is she able to compose and perform.

At larger scales, human technology, culture, and society all fundamentally depend on energy. Energy is required to produce jet skis and cell phones—jet skis have combustion engines

PHASE III: TERRAFORMERS

that require energy-dense fuels and cell towers are energetically expensive to operate.

Cities great and small require vast amounts of energy, primarily in the form of fossil fuels, to endure and grow, allowing them to become centers for human creativity, innovation, and economic activity.

11,000-Watt Lives

How then can we understand, in a unified way, all these different uses of energy by humans, much less other species? In the 1930s, Max Kleiber, a Swiss agricultural biologist, observed that, across mammal species, from shrews to elephants, the energy required to maintain basic metabolic function is closely correlated with an organism's body size.

Given that the average contemporary human consumes as much as thirty times more than a preindustrial human, our effective global population from an energetic point of view is closer to 210 billion rather than our planet's current seven billion humans.

Inspired by these patterns, biologists and physicists working together at the Santa Fe Institute in the 1990s and 2000s developed a mathematical theory to explain Kleiber's observation. This framework successfully predicts many aspects of the life and energy use of individual species, from how long an organism lives on average to the age at which it weans its offspring. It also predicts features of groups of species, such as how many more small species than large species there are in an ecological community.

Chapter 24: Are Humans Truly Unique?

As with other species, predictions can be made and evaluated for humans. Humans live about as long as the theory predicts for an organism of our size. We wean our offspring at the predicted age. We reproduce at close to the predicted rate. And so on.

It turns out that across many such life-history traits, there is nothing particularly unique about the basic biology of the human species—we fit right where we should on the body-size continuum.

But things get interesting when we look at the energy consumption of humans. Our basic metabolic rate, as predicted by our body size, is about 100 watts—the energy demand of an incandescent light bulb. That's about what you'd expect given our body size, according to the theory.

But in the modern industrialized world, the energy we actually consume, collectively, to fuel our lives—to do things such as construct roads and buildings, fly airplanes, drive cars, and harvest and refrigerate food—is closer to 11,000 watts, on average, for someone living in the United States.

In short, people in the US consume about 110 times more energy to function in an industrialized economy than predicted for an organism with our body size. The global human average, 3,000 watts, is thirty times greater than predicted. And that does make us unique; no other species on the planet uses close to this much energy to fuel their lives. What's more, given that the average contemporary human consumes as much as thirty times more than a preindustrial human, our effective global population from an energetic point of view is closer to 210 billion rather than our planet's current seven billion humans. More on that later.

Food Networks

Another way of thinking about human uniqueness is how we humans meet our fundamental energy needs.

PHASE III: TERRAFORMERS

Plants fuel their metabolic activity by converting energy from the sun and inorganic material from the soil, and most other species do it by consuming other organisms, including plants. As a result, all species are embedded in complex food webs.

Food webs are a type of ecological network representing the myriad feeding interactions among species as they all eat and are eaten. Researchers at the Santa Fe Institute and elsewhere have discovered ways to study networks of all kinds mathematically: transportation networks that connect cities by plane, train, and automobile; interactions among the proteins within a cell; and the conflict patterns of primates or the sexual relations of teenagers, for example.

Scientists are just beginning to understand how humans fit into food webs and how they compare to other species—in particular, the roles they play as predators. For example, humans migrated to the Aleutian Islands of Alaska several thousand years ago. The Aleut people who settled on the islands were hunter-gatherers who relied almost entirely on intertidal and marine species for food. They had a bounty of seafood at their fingertips, and they took advantage of it.

Of the hundreds of marine species close to shore, humans hunted, gathered, and ate almost a quarter of them, feeding on everything from algae and shellfish to fishes, birds, and marine mammals.

Network analyses of the feeding relationships among all species in this complex food web allow us to understand how humans fit and just how unique we are. Here's what we find:

Compared to other species in the region, the Aleut had the most diverse diets, which means they were not just generalist feeders—they were "supergeneralists." Of the other species in the web, only Pacific cod feeds as generally as humans. Most species in the web feed on fewer than ten species, compared to more than 120 eaten by humans.

Humans were also one of the most omnivorous species, feeding on everything from algae at the base of the food web to sea lions at the top of the web—and everything in between. Thus, the Aleut hunter-gatherers do seem to play unique roles in this marine food web.

What Else is on the Menu?

Once we start looking across food webs, however, it becomes clear that every web has a supergeneralist—something that is eating about a quarter of the species available to it.

For example, raccoons are supergeneralists. They feed omnivorously and opportunistically on fruits, nuts, and grains; a wide range of earthworms and insects; frogs, lizards, and snakes; small mammals; turtle and bird eggs; crayfish and other aquatic invertebrates; and fish. In urban areas, their diet expands even further to include roadkill, dog and cat food, and garbage.

Most species in the web feed on fewer than ten species, compared to more than 120 eaten by humans.

The identity of the supergeneralist changes, but one or two species always play that role in every food web. In this sense, the role of supergeneralist is not unique to humans.

Further, the Aleut people were similar to other predators in another way: given the wide variety of species they could feed on, they would switch their focus and effort from one prey species to another depending on conditions and availability. When it was

PHASE III: TERRAFORMERS

sunny and calm, for example, they would use kayaks to hunt sea lions. When the weather was stormy and the waters choppy, they would gather shellfish in the intertidal area, first focusing on easy-to-gather, large-bodied species, then shifting to smaller-bodied shellfish as the larger ones disappeared.

This "prey-switching" behavior is common to generalist predators and turns out to be ecologically stabilizing for the whole food web. As generalists turn their attention to different prey species, the species that were getting depleted get a chance to recover, reducing the risk of extinction.

Whether we look at ancient or modern people, *Homo sapiens* has always been extremely adaptable. Preindustrial humans had to make do with what was in their environment that they could forage for or grow, and some groups had more specialized diets than others. Likewise, today's humans embrace everything from narrow, locally sourced vegan diets to the highly general diets of the Anthony Bourdains of the world, who constantly seek out new food experiences from across the globe.

> **Put another way, by these measures, human uniqueness is tied to the industrialized societies we have created, not to our fundamental biology.**

Thus, what is uniquely human is the extreme scope and variability of our diet. But there's a new problem: in the industrialized modern world, humans don't always prey switch.

For example, as the bluefin tuna becomes harder to find due to overfishing, its economic feedback cycle value as one of the most prized sushi fish goes up, spurring more fishing, which drives the price up, and so on. Not only does this drive bluefin tuna toward extinction, it can potentially place other species in the marine food web at risk by interrupting the many interactions that connect species in food webs.

Modern Problems

Today, we humans consume a great deal more energy than our basic biology requires, metabolically speaking, and modern economic forces disrupt our ecologically stabilizing prey-switching behavior.

Put another way, by these measures, human uniqueness is tied to the industrialized societies we have created, not to our fundamental biology.

Understanding the roles humans play in these complex culture–environment–energy systems helps us assess the impacts of our actions in new ways. Identifying what makes us unique as a species points us to better ways of thinking about problems of sustainability, biological diversity, and other complex modern problems—and perhaps helps us get back in touch with our inner human.

ENGINEERED SOCIETIES

*Jessica C. Flack, SFI, and
Manfred D. Laubichler, Arizona State University
Christian Science Monitor, January 7, 2016*

In the American Southwest, in a remote canyon between Albuquerque and Farmington, New Mexico, lie the ruins of the cultural hub of the ancient pueblo peoples who populated the region roughly between AD 900 and 1150.

Chaco Canyon was the architectural and social masterpiece of its time, the region's center of trade, religion, and social organization. Its buildings were the largest in North America until the nineteenth century. Some appear to have been constructed so as to be aligned with solar and lunar cycles. A system of symmetrically radiating roads connected Chaco with the rest of the region.

Chaco was the result of decades or centuries of planning and building, and many of Chaco's features go well beyond those functionally related to survival. What purpose do these features serve, if not simply to provide shelter and security for Chaco's inhabitants?

To scholars, it's clear Chaco's design played a central role in setting up, maintaining, and reinforcing the complex social organization of the peoples who constructed it.

Human history is full of similar examples. The Balinese water temple system that emerged in the ninth century features iconic, stylized monuments and evolved rituals that optimize planting cycles and water distribution. The opaque voting protocol invented by Venetian families in the 1500s helped ensure tamper-free elections of their doge.

Humans have been attempting to engineer social outcomes presumably since language evolved and made feasible

PHASE III: TERRAFORMERS

the coordination of many individuals. It began when groups of hunter-gatherers decreed the first rules of social interaction, was advanced when the first agricultural societies set down regulations for water use and distribution, and has now expanded in contemporary society to our online behavior.

Reactive Social Engineering

A trained mechanical or electrical engineer might balk at using the term *engineering* to describe attempts to orchestrate social outcomes. Social engineering has never been a precise endeavor—nothing like designing a cell phone tower or constructing a water wheel for milling grain.

There are no meticulously drawn blueprints for social systems, and no objective tests that would allow us to predict the success of interventions. In fact, the history of human social engineering, right up until the present, is based largely on human intuition about the few directly attributable causes of a problem and how to adjust them.

> Scientific models that seek to predict the consequences of human actions with some reasonable accuracy—such as game theoretical models of economic behavior—for the most part ignore human individuality in favor of aggregated outcomes.

As a result, attempts at social control have been mostly reactive responses to the consequences of previous actions and decisions. If we think crime rates are too high, we invest in police and prisons.

Unfortunately, as the US incarceration rate attests, such ham-handed social solutions often backfire.

Is proactive social engineering desirable? Is it even possible to effect the outcomes we want in a medium as complex and uncertain as human social behavior?

The Cusp of Understanding

One reason for this lack of precision—assuming for the moment that social engineering could be deployed to advance the common good—has been a lack of detailed data about individual human behavior. Scientific models that seek to predict the consequences of human actions with some reasonable accuracy—such as game theoretical models of economic behavior—for the most part ignore human individuality in favor of aggregated outcomes.

But this is changing. With digital technology it is now possible, for the first time in human history, to track individual behavior and interactions, empirically study the behavior of humans in groups, and quantify the process of collective decision-making using such advances as geotagging, eye-movement tracking, and data mining of social media activities.

Currently, much of human social data collection is occurring online and is controlled by private companies. According to social media marketing strategist Jeff Bullas, as of 2014, 72 percent of internet users in the US and 64 percent of users worldwide use social media. Facebook has 1.55 billion monthly users, up from one million only ten years ago. The average American user spends more than one quarter of every online hour on social media, and almost 50 percent of Americans say Facebook is the number-one influencer of their purchases. Google+ has been around for only four years and already there are one billion Google+ enabled accounts.

Consequently, Google and Facebook are storehouses of detailed data on the minutia of human behavior, and they certainly are

PHASE III: TERRAFORMERS

experimenting with new kinds of social engineering, for better or for worse. In a controversial 2014 study published in the *Proceedings of the National Academy of Sciences*, scientists, in collaboration with Facebook, manipulated user feeds to study how negative and positive emotion spreads over social networks. In 2012, in a study published in the journal *Nature*, Facebook studied the effect of its political mobilization messages on real-world voting.

Without a doubt, corporate or government control of these data is a major privacy concern. But within the data (if appropriately anonymized) is immense potential for gaining fine-grained insights into social patterns and designs as, increasingly, people from all walks of life are living online lives.

Synced Multidata

Syncing time, location, and other data collected in this digiverse with all that goes on in the material world is not far off, and such "time series" data come with even greater potential. It is already possible, for example, to track an individual's movements via cell phone and, increasingly, through image capture coupled to image tagging.

One estimate of the number of video cameras used for surveillance worldwide is on the order of 100 million. By 2008, London—supposedly the most monitored city in the world—was estimated to have one camera for every fourteen people.

All your online activities synced with the world of bus schedules, stock markets, and weather . . . what's next? The data sets of the near future will contain systematically sampled data on a much wider range of behaviors at many complementary levels of analysis.

Proactive Social Engineering

What will we do with all this information? Is it indeed possible to deploy all of it for useful good?

Chapter 25: Engineered Societies

Machine-learning approaches that can find patterns and correlations in vast data sets already make decent predictions about your movie preferences, the style of clothes you are likely to purchase, whether you will comply with a doctor's orders, and the like.

But this is not yet *understanding*. Skeptics argue that these correlations will not help with designing social outcomes—that we will be no closer to engineering utopia, assuming we could agree on what might constitute utopia, than we were before.

Jorge Luis Borges captured this view in his famous short story, "On Exactitude in Science":

> In that Empire, the Art of Cartography attained such Perfection that the map of a single Province occupied the entirety of a City, and the map of the Empire, the entirety of a Province. In time, those Unconscionable Maps no longer satisfied, and the Cartographers Guilds struck a Map of the Empire whose size was that of the Empire, and which coincided point for point with it. The following Generations, who were not so fond of the Study of Cartography as their Forebears had been, saw that that vast map was Useless . . .

But Borges in some ways got it wrong. He is correct that detailed maps cannot be the goal—but they can serve as rigorous starting points. The behavioral maps we build with the vast data that are now being collected will allow us to find and quantify the hidden regularities in our social interactions—regularities we may think we understand but have never measured. Why is religious fundamentalism on the rise across the globe? How is it connected to income inequality, educational achievements, disease patterns, environmental degradation, and climate change? Why do we fear some people (xenophobia) and embrace others (xenophilia)? What factors, for individuals and groups, determine whether we perceive a foreigner as a threat or an opportunity?

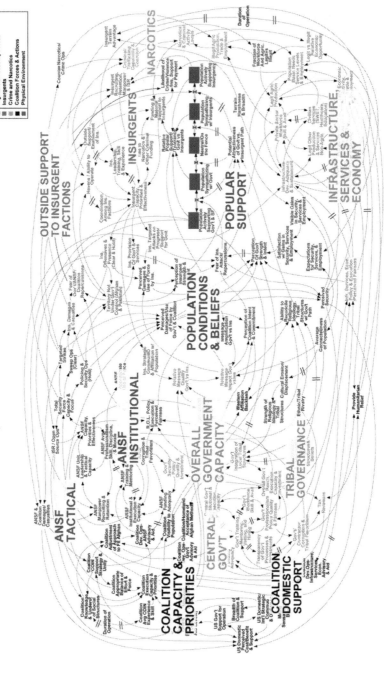

FIGURE 1. The "famous" causal-loop diagram shown to General Stanley McChrystal in 2009. PA Consulting Group

With an understanding of these regularities in hand, we will be in a position to infer the rules and strategies that humans use to guide their decision-making, and with this information build testable, predictive simulations of social outcomes at the societal level.

Such simulations would also allow us to test alternative futures—futures with both desirable and undesirable outcomes. These in turn would inform policy making (a.k.a. social engineering) that would be, for the first time in human history, empirically based.

This is complex systems science at its finest—reducing seemingly irreducible social complexity, through careful consideration of the data, to an elegant, compressed, predictive, and biologically and socially realistic theory of society.

As we develop this theory we will come to understand what factors matter. Which data can be ignored? How much of a role is there for chance and randomness? And, hence, what we can change to large and small effect at the individual and collective levels?

We may one day even be able to produce a rigorous, quantitative version of the now-infamous "causal loop diagram" (a.k.a. "spaghetti diagram") produced in 2009 by a consulting group for the US Army to explain counterinsurgency dynamics in Afghanistan (fig. 1).

Famously, when shown that one ponderous slide during a briefing in Kabul, the man in charge of the military operation at the time, General Stanley McChrystal, remarked: "When we understand that slide, we will have won the war!"

General McChrystal saw the futility in seeking understanding through wanton oversimplification of human social dynamics. Perhaps, though, by studying these dynamics empirically and in detail, we'll understand them well enough to not simply win wars but to avoid some of them altogether.

WHY PEOPLE BECOME TERRORISTS

Mirta Galešić, SFI
Christian Science Monitor, January 21, 2016

In the aftermath of any terrorist act, our instinct is to try to make sense of the brutality by assigning a person's or a group's violent radicalization to one or two probable causes: religious extremism and economic disparity, for example.

If we could only find a simple cause, our thinking goes, we might find a simple solution—blocking the influx of potentially dangerous ideas and people, or sending more military power to excise terrorist strongholds.

Such simplifications help us cope and make for powerful political messages, but it's clear that no single factor can explain violent radicalization. Inevitably, such narrow-sighted reactions only make the problem evolve into a different, and potentially more dangerous, beast.

Welcome to complex social systems, where the interactions of many individuals immersed in particular socioeconomic circumstances lead to the emergence of sometimes surprising social phenomena, from fashion trends to political movements, from conspiracy theories to financial crises, and from religious rituals to jihads. Complexity science can help us understand the underlying social systems from which these problems emerge; we should let it guide us in developing a thoughtful, science-based policy for dealing with them.

What Drives a Terrorist?

Decades of research have, thus far, not revealed any common psychopathological symptoms among terrorists. They appear to want some of the same things most of us want: recognition from

PHASE III: TERRAFORMERS

their peers and communities and better lives for the people they care about.

Being a devout member of a particular religion is not a good predictor of violence, either. Many deadly terrorist acts are unrelated to religious ideology. In recent years, Islamic values have been used to promote warmongering ideas across cultural and national borders by invoking the authority of widely revered holy figures and scriptures. But Islam is not terrorism and terrorism is not Islam.

Nor is the tactic of co-opting broadly accepted values or authorities to promote a certain idea unique to the current wave of "Islamic terrorism." Politicians and interest groups routinely build on shared values such as freedom, family values, or religious doctrines to promote their agendas.

Economic circumstances are an important but not paramount contributor to violent radicalization in the West. Poverty and limited job opportunities do cripple one's chances of achieving life goals, and discrimination is a real problem for many minorities. But very few poor or discriminated-against people become terrorists, and a significant number of terrorists have led stable middle-class lives.

Youth on the Edge

So what can we say are the genuine markers of violent radicalization in the West today? As noted by anthropologist Scott Atran, political scientist Olivier Roy, and other prominent terrorism scholars, three consistent characteristics are common to recent perpetrators of terror, and these hold for past terrorist movements as well:

1. Terrorists tend to be young, rarely older than thirty and usually in their early twenties.
2. Terrorists feel resentment toward mainstream society because of perceived or real injustices they have experienced, and they often feel frustrated that they are unable to obtain justice.

Chapter 26: Why People Become Terrorists

3. Before they became violent, most terrorists spent time in close contact with like-minded peers or charismatic leaders.

In essence, we are dealing with disgruntled young people who do unexpected or dangerous things under the social influence of their peers and role models—a.k.a. the common adolescent. And yet, only a small fraction of disgruntled young people stumble into behaviors that are significantly more dangerous than making severe fashion statements or partying excessively.

To understand how a seemingly "normal" person can become a terrorist, we need to understand not only how individuals are wired but also how people are wired together.

Social Animals

Social science suggests that all of us are defined by those with whom we hang out in at least three key ways:

1. The people around us influence how we perceive the global society. In other words, we use our own social milieu to make inferences about how people we don't know live their lives. But this may backfire when we live in homogeneous social environments and rarely meet people living in different circumstances. English psychologist Rael Dawtry and his colleagues have shown, for example, that people who live in richer neighborhoods perceive income distribution in the general population as more fair than it really is. Consequently, they are less likely to support policies aimed at reducing the gap between rich and poor.

2. Our social circles have a strong influence on our own beliefs and behaviors. In the 1950s, conformity experiments by social psychologist Solomon Asch showed that some people will disregard objective facts if everyone else opposes them. More recently, social scientists Nicholas Christakis and James Fowler have found that people tend to mirror their social contacts when it comes to

PHASE III: TERRAFORMERS

dietary habits, smoking and drinking, and even emotions. This similarity reflects the underlying processes of social influence and the tendency to befriend like-minded people.

3. We feel good when others agree with us and bad when they disagree. All of us have experienced unpleasant feelings arising when someone expresses opinions with which we profoundly disagree. These feelings of anger and frustration were labeled *cognitive dissonance* by the psychologist Leon Festinger. They motivate us to try to influence disagreeing others, and if that fails, stop communicating with them altogether.

What I've described so far are all useful aspects of human sociality that help us to learn from each other, form efficiently functioning groups, and stand as one against a common enemy when necessary. In fact, many scientists agree that our ability to learn from and cooperate with others may be the cornerstone of the spectacular success of our species. But sometimes our advanced sociality helps spread beliefs that are harmful to a society.

How Beliefs Form and Change

Complexity science—by its nature an amalgam of fields—permits us to pull together techniques and tools from cognitive and social psychology, network theory, anthropology, political science, economics, statistical physics, computation, and more for deeper quantitative insights into social systems than any single field can offer. This can help us study how our social-cognitive processes interact with and shape our social networks and our broader economic and political circumstances, and gives us hope of explaining when and why harmful beliefs spread.

Consider a social-cognitive process called *social sampling*. Our social environments—dominated by our social contacts and the media—constantly serve up a menu of beliefs and behaviors we might try. Most are harmless or beneficial, such as a new fashion fad or an improved technology. But some, such as excessive nationalism or denigrating entire ethnic or religious groups, can be very dangerous. Some people, depending on their social environments, and in particular on the diversity of their social contacts and media sources, will be exposed to such dangerous ideas more frequently than others.

A second important process, belief updating, determines whether we will accept a novel idea coming from our social environments. Its outcome depends on the strength of our own views, formed by a lifetime of experiencing our own particular societal circumstances. But it also depends on views of the social contacts we most respect, feel personally close to, or wish to impress. Hence, we are more likely to accept a dangerous idea if it aligns with our own experiences and is supported by the people we value.

A third, significant social-cognitive process is network updating. An easy way to avoid the unpleasant feelings of cognitive dissonance is to not discuss our beliefs with contacts who disagree, or to cease socializing with them altogether. As a result, our social circles become more homogeneous, and beliefs our former peers might have considered extreme are no longer challenged by anyone important to us.

None of these processes in isolation is dangerous, *per se*. But taken together, in some societal circumstances, they can feed a spiral leading some individuals toward increasingly extreme and sometimes dangerous beliefs.

Systemic Understanding

The social-cognitive processes I have described are shaped by millennia of human biological and cultural evolution. Suppressing them would not only be very difficult, it would also be counterproductive.

PHASE III: TERRAFORMERS

In most parts of our lives, they are essential and desirable components of our exceptionally well-developed sociality.

What we might change are the inputs to these processes: the content of individuals' social samples, the personal experiences shaping their views, the diversity of their social circles, and the ease of making and breaking new social connections. In turn, these processes will feed back into and act upon our society. Understanding this cycle can help us design interventions that not only affect one of its parts but have positive feedback effects throughout the system.

Consider, for example, interventions targeting the input to social sampling processes, such as paying attention to the information available to young people on social media or crowding out dangerous ideas by suggesting alternative ways for young people to join exciting and meaningful causes and achieve peer respect. By altering the ingredients of the idea menu they are exposed to, we might, in turn, minimize the dangerous inputs to the processes of belief and network updating.

Interventions targeting belief-updating processes include exposing chauvinistic and violent ideas disguised as legitimate spiritual or patriotic values. In many cases, these interventions would need to come from within the affected communities—and be championed by their internal role models.

Equally important for belief-updating processes is improving the economic and political circumstances that limit young people's opportunities to achieve life goals in nonviolent ways. By eliminating some of the justifications for the acceptance of violent ideas and providing peaceful paths to self-realization, such interventions also shape the content of people's social samples.

Interventions that target network updating, such as efforts to avoid alienating youth from mainstream society or to identify those in the process of withdrawing, will in turn affect inputs into their social sampling and belief-updating processes.

And so on. The social cognitive–environment feedback cycle offers many potential points of intervention.

None of these ideas is new, of course. All have been tried or are being implemented in different countries and communities. Nor are they, in isolation, the final solutions to the problem of violent radicalization.

Complexity science can help us take in the bigger, messier picture. We can build quantitative, empirically grounded computer models—simulations—of how these interacting processes interact with and influence people and the societies in which they are embedded. Within these models, we can try interventions and see if they succeed or fail. We can design ensembles of strategies with synergistic effects, then evaluate them and their effects on the system as a whole rather than as isolated tactics.

Ultimately, by thinking more deeply about the many ways we influence each other within our complex social circumstances, we might find ourselves better able to comprehend and cope with terrorism—and the other emergent social challenges that keep vexing us, including the debate for and against immigration, the gun-control conundrum, the ongoing redefinition of gender roles and anger-fed political movements.

BEEHIVES AND VOTING BOOTHS

John H. Miller, Carnegie Mellon University
Christian Science Monitor, February 11, 2016

During every US presidential primary season, we watch as the political fortunes of individual candidates rise and fall, seemingly without regard to whether a candidate has the skills, character, or ideological foundations to govern productively. To the casual observer, the process looks like unmitigated chaos. Inevitably, this political circus results in nominees and, after the general election, a winner—someone who serves as our president for four years—even if we don't all agree with the choice.

Messy, yes. But when you think about it, our political system is a remarkable outcome of human sociality. We intentionally create social chaos to achieve social order. What forces underlie this process? Perhaps most important, does it—can it—result in wise choices?

How Order Emerges

The study of complex systems, like all of science, is a search for order. Traditionally, science seeks order by understanding the simplest parts of a system. How does a single gas particle behave given a certain temperature? Which gene in our DNA determines eye color? Scientists then try to develop theories that explain more general observations based on their detailed understanding of the individual parts.

Complex systems science is different. It seeks order by understanding how simple parts, interacting together and perhaps adapting to one another, create an entirely new whole. The collective outcomes of complex systems can be surprising because the parts often don't add up as expected.

PHASE III: TERRAFORMERS

In his book *Wealth of Nations*, published in 1776, philosopher and economist Adam Smith noted how an individual selfishly seeking his own security "is in this, as in many other cases, led by an invisible hand to promote an end which was no part of his intention." Remarkably, while the study of economics has developed a sophisticated theoretical apparatus over the past 240 years, there are incredible gaps in economic theory that even today are filled by invoking "invisible hands" guiding the behaviors of economic agents.

Famously, the 2008 financial collapse—one of the most important economic events in the modern era—was widely unanticipated by economists. Equally as worrisome, once the crisis emerged, the prevailing theories yielded a dearth of prescriptive advice. It was as if a geologist happening upon the rim of the Grand Canyon was able only to proclaim: "Something happened here."

> At the heart of the complex system embodied by a beehive is the mystery of how a collection of many thousands of bees operates without apparent control, yet with an efficiency that would be the envy of most industries or governments.

The 2008 crisis embraced all of the seven deadly sins, from gluttonous banks to greedy mortgage brokers. But the biggest failure—our collective blind spot—was failing to appreciate the big, interconnected picture. While economists and policy makers were well equipped to understand and control the individual parts that contributed to the crisis, they were unable to comprehend how those parts added up to the whole.

In the case of the financial crisis, various positive feedback loops linked the different parts of the system, and the same forces that amplified the system on the way up (much to the delight and profit of all involved) accelerated its demise on the way down. The real danger was systemic.

Birds, Brains, and Bees

Over the last couple of decades, we have begun to develop new observations and techniques that allow us to understand how systems emerge from pieces. Good examples are often found in nature.

Flocks of starlings, known as *murmurations*, undulate in the sky like beautiful ghosts. Such seemingly coordinated collective behavior arises when individual birds follow a few simple rules: stay close, but not too close, to your neighbors and fly roughly in a similar direction and at a similar speed.

Likewise, the individual neurons in your brain respond in simple ways to signals from their local connections. There's still a lot we don't know about brains, but neurologists have demonstrated that decentralized mechanisms like signal averaging—essentially, your neurons voting based on the preponderance of signals near them—set up a self-reinforcing feedback loop that probably has a lot to do with how we form thoughts.

For a wonderful example of decentralized decision-making in nature, consider a beehive. Queen bees, their royal titles notwithstanding, are little more than egg-laying machines. Indeed, no centralized hive governance system has ever been discovered.

At the heart of the complex system embodied by a beehive is the mystery of how a collection of many thousands of bees operates without apparent control, yet with an efficiency that would be the envy of most industries or governments.

PHASE III: TERRAFORMERS

The behavior of each bee in a colony is governed by a simple set of rules that, through interactions with the environment and the other bees, leads to an end that was no part of any individual bee's intention, allowing the colony as a whole to thrive—as if by an invisible wing. Put another way, bees are relatively dumb. Beehives are remarkably smart.

Making the Best Choice

One example of bees making smart decisions arises when they need to find a new home. Scout bees in the swarm are encoded with a simple exploratory behavior that motivates them to search for potential hive sites.

When a scout finds a potential site, she returns to the swarm and does a "waggle" dance—which looks a bit like human twerking—to advertise the new site's location to the other scouts. Her dance will often motivate other scouts to check out the site and return to dance on their own. Key to this process is a positive feedback loop in which the better the apparent quality of the new site to the scout, the longer she dances.

Over time, a variety of locations are explored, with the prospects of each site rising and falling as dancers come and go. Eventually, when a sufficient quorum of scouts forms at one of the locations, it becomes the final choice. Remarkably, this process allows the bees to find a high-quality location in a timely manner without any kind of central authority, as it is only the local actions and observations of the individual scout bees that drive this system.

What's most surprising is this: both theoretical models and field experiments show that bees tend to make the best choice from among the available options. The positive feedback mechanism of scouting, communicating, and verifying in increasingly greater numbers causes seemingly better options to be explored more intensively. As long as the quorum is large enough, the vagaries of the initial search process diminish and a good final choice results.

Benefiting from Buzz

Bees offer an interesting parallel to our political system as well. Candidates in presidential primaries are a lot like potential hive locations, and media attention, campaign donations, and even yard signs and bumper stickers constitute the political "dance." As a candidate's political fortunes rise or fall, he or she garners more or less attention and evaluation—essentially "buzz"—creating the necessary feedback loop.

As long as this feedback loop is tied roughly to the true quality of each candidate, as with the new beehive location, the apparent chaos of the system tends to result in a nicely ordered final choice.

Thus, while media attention driven by particular political ideologies or outsized monetary support may encumber the primary system, having multiple debates and requiring candidates to garner sufficient support from a series of delegate elections over time should allow the best candidate to emerge from among the pack.

Complex Collective Problems

Complexity science, a relatively recent arrival on the scientific landscape, offers a number of new ways for exploring what might be our last remaining scientific frontier: how unexpected system-wide phenomena emerge from parts.

We can, for example, use the ever-increasing power of the computer to form agent-based models of individual, computerized traders haggling in an artificial market to gain new insights into how prices arise (perhaps allowing us to wave goodbye to Mr. Smith's invisible hand), or how bubbles form or markets collapse.

PHASE III: TERRAFORMERS

Understanding the structure of interconnected systems using network science allows us to analyze the various loan obligations made across the banking system and identify conditions in which a small number of failures can bring the entire system down.

> **Complexity is at the core of most of the major challenges confronting humanity—climate change, financial collapse, ecosystem survival, inequality, terrorism, and disease. If we understand complexity well enough, perhaps, the same complexity that creates these problems can help us choose a politician able enough to address them.**

Simple programs adapting to each other inside a computer give us insights into how cooperation can emerge in a social system, even when each individual would be better off by not cooperating.

Studying the complex social and agricultural rituals Balinese rice farmers have developed can give us insights into how they have been able to sustainably farm their rice terraces for centuries, even though water is chronically scarce enough to threaten a complete breakdown of the system.

Computational models showing how individual families' (even mild) preferences to live in a neighborhood with families like them can quickly devolve into a highly segregated cityscape offer insights into wealth inequality, race relations, and urban planning.

We can study the apparent chaos of the beehive or anthill to gain a better understanding of the order that emerges in our governments—and when central control or decentralization makes the most sense.

Complexity is at the core of most of the major challenges confronting humanity—climate change, financial collapse, ecosystem survival, inequality, terrorism, and disease. If we understand complexity well enough, perhaps, the same complexity that creates these problems can help us choose a politician able enough to address them. ⚡

THE SOURCE CODE OF POLITICAL POWER

Simon DeDeo, Indiana University
Christian Science Monitor, March 24, 2016

People search for information with Google and Yahoo, but they often find it on Wikipedia, the sixth most-visited site on the internet. Everyone brings their questions there: students use Wikipedia to cram for exams; journalists, to check their sources; scientists, to broaden their view of a field.

Famously, Wikipedia isn't a well-planned operation. Its salaried employees are massively outnumbered by tens of thousands of "editors" who, attracted by Wikipedia's vision or irritated by its inaccuracies, take it upon themselves to contribute. And contribute they do: researchers at the University of California, Berkeley and the University of Minnesota estimate that volunteer editors have invested more than forty million hours of labor into Wikipedia, comparable to the effort required to build the underwater tunnel between Britain and France.

These contributing editors not only write articles, they also argue with each other. How should they describe controversial issues or track down hoaxes and errors? What should they do with ill-intentioned or chronically ill-behaved editors? How should they punctuate the movie title *Star Trek Into Darkness*?

Debates on these questions spill out well beyond any particular conflict, as editors write essays, propose policies, and build support for guiding principles—some might say rationalizations—to bring order to the chaos. What happens when they try?

PHASE III: TERRAFORMERS

More is Different

Questions such as these are more than just occasions to tell stories: they're also questions that scientists try to answer. In the twentieth century, social scientists—impressed by the success of physics and chemistry—tried to provide general answers by writing down mathematical laws to describe and explain social behaviors. When they did so, they imagined a world in which everyone was the same. This assumption made calculations possible, and occasionally it even predicted what people would do.

But these tools lagged behind what astute observers already knew: when people interact with each other, unexpected, unanticipated patterns emerge. Society is not one man writ large. When enough people gather together, what they create has its own, autonomous life.

> **An unexpected analogy to what we see in Wikipedia is to the Talmud, the book of Jewish law, whose basic texts underwent a similar pattern of growth through commentary—albeit over the course of centuries, not years.**

The Nobel laureate (and Santa Fe Institute affiliate) Phil Anderson made this principle a fundamental law of complex systems, succinctly expressed in the title of his famous article, "More Is Different." Anderson, a physicist, had won the Nobel Prize for the study of how crystals conduct electricity when they have imperfections. His essay went far beyond physics, however—and it inspired many of us to strive to understand the imperfect crystals of society, made of human minds.

It was in that spirit that my Indiana University colleague Bradi Heaberlin and I trawled the hidden side of Wikipedia—a human society in which every editor's moves are documented down to the second.

We drew up nearly 2,000 pages, a dense network of interconnected policies, expectations, and social norms: accounts of how a good editor ought to behave. As it was added to and woven together over time, this "norm network" tracked the hopes and fears and struggles of tens of thousands of users since the project began in 2001.

Ecosystems of Ideas

When we turned our mathematical tools on this hidden world, we didn't find a handbook or a *For Dummies* guide, the product of a single mind. She and I found something far more interesting: an evolving ecosystem of ideas.

Some pages urged users to be civil or to be neutral, for example, while others, written later, tried to understand what being civil, or being neutral, really meant. Some pages were concerned with truth, others with proper formatting (in case you're wondering: Wikipedia is formally neutral on the Oxford comma). Some talked about the importance of being polite, but others warned about how a preoccupation with politeness can undermine excellence.

The patterns that emerged also changed over time. As Wikipedia's network grew, halos of commentary emerged to feed off the core policies. An unexpected analogy to what we see in Wikipedia is to the Talmud, the book of Jewish law, whose basic texts underwent a similar pattern of growth through commentary—albeit over the course of centuries, not years.

In Wikipedia, commentary often signaled the authority of what it commented upon: the more commented on a page, the more often

PHASE III: TERRAFORMERS

it was accessed, edited, and talked about. At the same time, however, commentary had a second, somewhat unexpected effect: it pulled the network apart. Debates on how to be neutral, for example, tended not to connect to debates on how to be civil. We could track the emergence of distinct communities of interpretation, as if watching the emergence of different forms of law over time.

Most interesting yet, we found a strong signal of what scientists call a founder effect: in many cases, the enduring norms were first written years before Wikipedia grew to fame, and when the population was a small fraction of what it would one day become. Just as the foundational precepts of the United States still govern us today and generate much commentary in the form of Supreme Court opinions, Wikipedia is governed by rules that originated among a small minority.

None of this was planned, of course: no superuser wired these networks together, or later pulled them apart. Instead, dynamic patterns emerged from the actions of thousands of individuals, each with their own idiosyncratic beliefs, working with and against each other to define what it meant to be a good Wikipedian.

And while a small minority may have drafted the rules, it was the actions of those who came later that gave them the status they enjoy today. None of this could have been predicted by studying the motivations of individuals in isolation: the patterns we see are a fundamentally social affair. More, as Anderson said, is different.

Rerunning the Tape of History

What could Wikipedia have been like if that early group had chosen differently? If the group who drafted the influential policy "Neutral Point of View," for example, had chosen a different title or changed their minds? Would Wikipedia have failed? Would it have just been different? Or would others have stepped in to write the very same thing?

Chapter 28: The Source Code of Political Power

The biologist Stephen J. Gould once asked this question about life itself: what would happen if we reran the tape of evolution? Basic problems in how societies change and adapt, thrive, or fail to prosper depend on the answer.

It's a question we ask about societies much larger and grander than Wikipedia. With our colleagues in political science, for example, we're now looking at the development of the speeches in the United States Congress. And our tools allow us to compare how the United States runs itself (or doesn't) to what happens today when new democracies emerge from authoritarian control.

Our goal is to make maps: not of the Earth but of the landscape of ideas that people use to govern themselves. And we benefit from a new wave of digitization that allows us to look back as well as forward.

Centuries-old records from the United Kingdom, for example, have allowed us to track how the courts of London managed crime over the course of decades. We can see the emergence of new patterns of talking about crime, as courtroom participants learned to talk about violence in increasingly distinctive ways, marking a desire to manage violence differently from property crimes and disorder. In fact, in eighteenth-century England, pickpocketing was a capital offense, and a thief could receive as harsh a punishment as a murderer. We see the creation of new "genres" of criminal justice, new ways to talk about the crimes people commit—a pattern that continues today.

We venture into worlds that the mathematically minded have rarely been able to go—to the other side of the English Channel, for example, where the French government put the archives of the French Revolution online. Now digitized by the libraries at Stanford University, these texts give us the chance to look in at one of the turning points of the modern world, as a small group—roughly the size of Wikipedia's early editing population—abolish feudalism,

PHASE III: TERRAFORMERS

slavery, and hereditary aristocracy; separate church and state; and draft the declaration of the rights of man.

Vive la Différence

On the broadest level, we see similar patterns in the history of London, the emergence of new democracies, and the evolution of Wikipedia: toward increasing complexity, refinement, and—often—bureaucracy. Systems make distinctions and create differences, whether they be between ideas, parties, or people. And, as our colleagues Tim Hitchcock and Robert Shoemaker find in their studies of London, these patterns coexist with the tactics ordinary people use, as they leverage big ideas to get on with their lives.

> Whether we look inside parliament houses or web servers, we see dynamism and change: new ideas and unexpected logics of development. Some we may find silly, outdated, or even abhorrent, others exciting, new, or perplexing.

It's exciting to see. But we're wary: even the best can stumble when they look on the very largest scales. In the early 1990s, the great political scientist Francis Fukuyama cast a learned gaze over the violence of the twentieth century. He declared that the millennia-long struggle for recognition—the innate human desire to be something in the eyes of others—had finally ended, with the invention of liberal democracy. One idea had triumphed. He titled his essay "The End of History." But history didn't end, and new ideas continue to appear. One need to look only as far as the internet, where its offspring social media has now democratized the seeking and granting of attention. Would Professor Fukuyama have come to a different conclusion having seen the Islamic State on Twitter?

Will my colleagues and I do any better? I'm sure that in a few decades our theories will be easy to criticize as blind, even prejudiced. But one mistake I think we will avoid is the idea that social worlds evolve toward a stationary state. Whether we look inside parliament houses or web servers, we see dynamism and change: new ideas and unexpected logics of development. Some we may find silly, outdated, or even abhorrent, others exciting, new, or perplexing.

When we study the complex patterns that emerge from human interaction, we find laws of invention, turmoil, and creation—not stability. For Americans watching the primary season, that's not hard to believe at all.

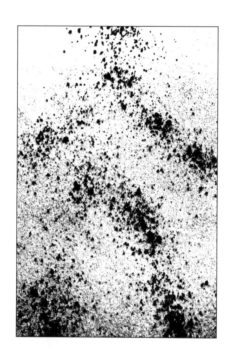

THE COMPLEX ECONOMICS OF SELF-INTEREST

Samuel Bowles, SFI
Christian Science Monitor, June 23, 2016

An email to me recalled the "exciting and stimulating times" that an old friend of mine had spent in the early 1950s as a staffer in the Executive Office of the President. "People worked long hours," he told me, "and felt compensated by the sense of accomplishment, and . . . personal importance. Regularly a Friday afternoon meeting would go on until 8 or 9, when the chairman would suggest resuming Saturday morning. Nobody demurred. We all knew it was important, and we were important . . ."

But then something changed.

"What happened when the president issued an order that anyone who worked on Saturday was to receive overtime pay . . . ? Saturday meetings virtually disappeared."

The emails were from Thomas Schelling, who half a century after he left the White House was awarded the Nobel Prize for convincing economists that their discipline should broaden its focus to include social interactions beyond markets. Things like segregated neighborhoods and business organizations and traffic jams and epidemics and information sharing on the internet.

The take-home message from Schelling's story—that incentives sometimes backfire—is familiar to psychologists. In a 2008 study, kids less than two years old avidly helped an adult retrieve an out-of-reach object in the absence of rewards. But if they were rewarded with a toy for helping the adult, the helping rate fell off by 40 percent. (I review dozens of similar experiments done by economists in my recent book, *The Moral Economy: Why Good Incentives Are No Substitute for Good Citizens*).[1]

---
[1] New Haven, CT: Yale University Press, 2016.

PHASE III: TERRAFORMERS

Emergent Morality

That incentives sometimes backfire, as Schelling discovered, is definitely a problem if you are an economist (I am one). Incentives, according to the dismal science, are the foundation of a well-ordered society. Adam Smith, considered by many to be the father of economics, said it well: "It is not from the benevolence of the butcher, the brewer, or the baker that we expect our dinner, but from their regard to their own interest."

And this, he explained in *The Wealth of Nations* (1776), is not altogether a bad thing: the butcher (along with the baker and other economic actors) "intends only his own gain, and he is in this, as in many other cases, led by an invisible hand to promote an end which was no part of his intention. Nor is it always worse for the society that it was no part of it. By pursuing his own interest he frequently promotes that of the society more effectually than when he really intends to promote it."

Smith's idea still resonates. In the aftermath of the stock market crash of 1987, the *New York Times* headlined an editorial "Ban Greed? No: Harness It," which continued: "Perhaps the most important idea here is the need to distinguish between motive and consequence. Derivative securities attract the greedy the way raw meat attracts piranhas. But so what? Private greed can lead to public good. The sensible goal for securities regulation is to channel selfish behavior, not thwart it."

The radical idea here is that shabby motives can be harnessed for elevated purposes. Thus, how well a society or economy works does not depend on a summing up of the quality of its citizens. What really matters is how they interact. In the language of complexity theory, how well an economy works is an emergent property of the interactions of the people making it up: it is something about the whole that cannot be inferred from the parts, or at least not by adding up the parts, or by any other simple rule.

Smith was among the first to use the idea (not the language) of emergent properties to understand the workings of the economy. (My Santa Fe Institute colleague John Miller, in chapter 27 of this book, had more to say about this.) But Smith must have missed something important. Economic self-interest may have put the beef, the beer, and the bread on the table, but it did not get Schelling and his colleagues to show up at the White House for Saturday meetings.

> How well a society or economy works does not depend on a summing up of the quality of its citizens. What really matters is how they interact.

Where did the classical economists go wrong? Neither Smith nor the great nineteenth-century economists who followed him made the mistake of thinking that people are in fact entirely selfish. Smith, in his other great book, *The Theory of Moral Sentiments*, held that "How selfish soever man may be supposed, there are evidently some principles in his nature that interest him in the fortunes of others, and render their happiness necessary to him, though he derives nothing from it except the pleasure of seeing it." John Stuart Mill, three-quarters of a century after Smith, termed the assumption of unmitigated self-interest "an arbitrary definition of man."

When More is Less

What Smith (and most economists since) missed is the possibility that moral, generous, or other socially beneficial behavior would be affected—perhaps adversely—by incentive-based policies designed to harness self-interest. To most economists the unspoken first principle is that incentives and morals are what is called *additively separable*. This mouthful of a term is from mathematics; it means that the effects of the one do not depend on the level of the other. When two things are additively separable, they are neither synergistic—each contributing positively to the effect of the other, like a duet being better than the separate parts—nor the opposite.

You have already seen where separability can go wrong. Offering Schelling and the White House staffers overtime pay for Saturday meetings was not simply another reason to show up, additional to their sense that what they were doing was important. Their public spiritedness was not separable from their self-interested regard for their own pay. The policy addressed to the latter appears to have diminished the former. The whole in this case was less than the sum of the parts.

What economics has missed is that adding an incentive—a fine or a bonus—may be subtracting something else, the individual's sense of responsibility, or obligation, or intrinsic pleasure.

This brings us back to complexity, a way of thinking in which the effects of things are rarely simply additive. Smith's invisible hand argument had the whole, namely, the dinner on the table, being greater than the sum of its parts, namely, the self-interested motives of those providing the food, who, Smith supposed, could not have cared less about the hungry family about to sit down.

But it can also go the other way. What economics has missed is that adding an incentive—a fine or a bonus—may be subtracting something else, the individual's sense of responsibility, or obligation, or intrinsic pleasure.

Psychologists Felix Warneken and Michael Tomasello, authors of the study about how rewards crowded out infants' intrinsic desire to help others, concluded: "Children have an initial inclination to help, but extrinsic rewards may diminish it. Socialization practices can thus build on these tendencies, working in concert rather than in conflict with children's natural predisposition to act altruistically."

This might be good advice for public policy, too.

WATER MANAGEMENT IS A WICKED PROBLEM, BUT NOT AN UNSOLVABLE ONE

Christa Brelsford, SFI and Arizona State University
Christian Science Monitor, July 13, 2016

Last summer, it was hard to miss news about California's drought, caused by the four driest years in the state's history. Its impact on California's economy in 2015 alone was estimated at $2.7 billion dollars and 21,000 jobs lost. Thanks to El Niño, this drought has eased some, but 42 percent of the state is still in a condition of extreme drought.

In 2007, there was a drought that didn't garner quite the same national attention: Atlanta, Georgia, was in a state of exceptional drought from September to December and came within a few months of running out of water. A large American metropolitan area running out of water almost certainly would have required driving in massive trucks of water every day just so that people could wash their hands, drink, and use the bathroom.

To address the impending disaster, Georgia's governor led several hundred people in prayer, without leading an official effort to reduce water demand or increase supply.

Unlike California's drought, Atlanta's was not unprecedented. Droughts of equivalent severity occur about once a decade, but the region's population had grown by 25 percent in the previous decade, and an increase in water-intensive agriculture was putting stress on the system even before dry weather limited Atlanta's supply.

Growing urban populations and increasing demand are likely to make water scarcity much more common, not just in Atlanta and California but across the US and in most arid parts of the world. What's worse is that these expectations do not take into account

PHASE III: TERRAFORMERS

the effects of continuing climate change, which will put additional stress on water supplies in many regions.

Unpredictable or Complex?

But we are not helpless observers. We can use ideas from complex systems science to find ways to improve the water supply system we have so that it better serves today's needs. First, it helps to think about what we control and what we don't.

> **Droughts often cause water scarcity, but the social and economic disruptions that follow from drought are a result of how water is managed, not because droughts can't be anticipated.**

Droughts are an expression of natural conditions: lower than usual precipitation, drier soil, and low stream flow. They result from normal and recurring weather variations. We can't predict when a drought will begin or end or how bad any given drought will be, but we do know from hundreds of years of meteorological records how often droughts of a given severity can be expected to occur.

Water scarcity, on the other hand, occurs when people want more water than is available. This distinction matters: droughts often cause water scarcity, but the social and economic disruptions that follow from drought are a result of how water is managed, not because droughts can't be anticipated.

This is not to say that water managers are doing their jobs badly. Our water system relies on close couplings between the hydrological and ecological contexts and the water distribution infrastructures we have built, as well as the social, economic, and legal institutions that have developed in this context. No single group or perspective can address all these features simultaneously.

This multifaceted complexity is why water allocation problems today are so wickedly difficult to solve. But solve them we must. Uncertainty over water access is inhibiting innovation, damaging ecosystems, and limiting economic growth.

Water is not Gold

If the system isn't working but it's no one's fault, how did we get here? The core ideas of the legal infrastructure that allocates water in the western United States were developed during the Gold Rush era of the mid-1800s and relied on legal precedents that determined allocation of mineral rights. This system gave ownership of the land and any minerals thereunder to the first person who invested the hard physical labor of digging for gold on that claim.

If a claim was abandoned, the next miner who started digging earned the right to any minerals he found there. This requirement, that a resource be used to maintain ownership, prevented miners from "staking their claim" to more land than they could reasonably work. It also provided a legal mechanism that allowed somebody else to use the resource if a miner disappeared or died. The two underlying legal principles that resulted from this system are known as *prior appropriation* and *beneficial use*.

But water is not gold. Unlike gold, water quickly moves from claim to claim through rivers and underground aquifers. This fluidity makes owning water much harder than owning gold. As a consequence, we define water use rights based on the expectation that there will be water in a stream, lake, or well during a given season if it has been there in the past.

When hydrological and human water use patterns are consistent over a period of many years, understanding the detailed hydrology

PHASE III: TERRAFORMERS

doesn't matter much, because water rights can be granted based on expected conditions. This is how the principle of prior appropriation is applied to water: if water is generally available in a place and time, the first claimant is granted the right to use water there as long as they maintain continuous use.

Crucially, once a water right has been established, any change in use from the right holder is permitted only if it doesn't damage other established users. If a pair of water users would like to arrange a water transfer between themselves, they need to demonstrate that this transfer won't harm any other users.

Consequences

Water transfer is why the Salton Sea in California still exists. In 1905, an engineering accident led to the Colorado River overflowing into irrigation canals and from there into a dry prehistoric lake bed; for almost two years thereafter, nearly the entire flow of the Colorado River was diverted, creating today's 350-square-mile Salton Sea.

> The complex system of water supply, demand, and use is evolving, but the legal system by which it is managed is locked rigidly in place.

The lake was quickly put to beneficial use, even though its only continuing water source was wasted irrigation runoff from those same irrigation canals. People drawn to the new inland lake quickly made the Salton Sea a tourist and recreational destination, and migratory birds and endangered fish have incorporated the lake into their ecological niches.

Chapter 30: Solving the Wicked Water Management Problem

In recent decades, thirsty cities like San Diego have looked to the Salton Sea as a water source, but based on the legal principle of prior appropriation and beneficial use, any claim they make is junior to the lake's established users.

Despite this, in 2003 the Imperial Irrigation District and San Diego agreed to a water transfer. San Diego would pay for efficiency improvements in Imperial's irrigation systems, and the city could then use the water saved to support its rapidly growing population.

But the agreement, which seemed like a win-win, presented a conundrum. By using more efficient irrigation technology, the Imperial Irrigation District would have reduced the agricultural runoff that had fed the Salton Sea. This reduction would have, in turn, damaged the lake's ecosystem and thus the users, who argued that they had been putting the ignored agricultural runoff to beneficial use for decades.

This, in turn, made it much more difficult for the irrigation district to invest in more efficient irrigation technology and sell the savings to a thirsty city. San Diego and Imperial's agreement did eventually get modified and approved, but only after a long legal battle with the lake's user groups.

Complex and tangled social–legal–economic circumstances like these are why water transfers are contentious, uncertain, and rare. Still, transfers are the only way to allow our allocation system to adapt to changing use patterns as the population continues to grow and new industries rise, fall, and evolve.

Put another way, the complex system of water supply, demand, and use is evolving, but the legal system by which it is managed is locked rigidly in place.

PHASE III: TERRAFORMERS

Where Might We Go from Here?

The doctrine of prior appropriation worked well while there was still unclaimed water available, but that era is over. Severe droughts like California's and increasing demand like Atlanta's, both ubiquitous features of our modern water climate, highlight the two faces of scarcity.

Thanks to increasing population, climate variability, and changing demand patterns, water scarcity will be a significant constraint in the western US for the foreseeable future. There is no new water to allocate, and so the water management task now is to make the best possible use of the water resources that are available.

Most conventional thinking about the water system ignores the deep complexities of water scarcity; as a society we think of taking shorter showers. Other oft-proposed solutions are big infrastructure projects to import water from somewhere else or investments in expensive water supply technologies such as desalination plants.

Water conservation and new technology will help us make better use of our resources. They are necessary components of any solution. But measures like these don't address the underlying cause of the water scarcity we face. One way or another, we need to figure out how to transfer water from one use to another, more critical use in a reliable and consistent way so that users not involved in the trade are not damaged by it, and so both agricultural and urban users can plan and make investments with reasonable confidence in the volume of water they'll be able to access—and the price they will have to pay to get it.

These problems are complex, but they are also solvable. Water management lies at the intersection of economic, legal, political, hydrological, climatological, ecological, agricultural, and engineered systems. It can be difficult for existing institutions to understand all the disparate perspectives. But we can use the frameworks of complex systems to begin to analyze water supply and demand.

Interdependency clearly plays a role in the systems we have and the systems we need to build, as does path dependence (in physics,

Chapter 30: Solving the Wicked Water Management Problem

a system whose state depends on the path taken to achieve it). The concept of scale helps us understand the spatial mismatch between the laws of physics that govern hydrology and the local, state, and federal laws that govern water allocation. Understanding the drivers and effects of institutional hysteresis (lag in response to forces) can help us break down impediments to better water supply systems. These are but a few examples of complex systems concepts that can help us think more deeply about water.

In turn, the knowledge that emerges from this deeper thinking can facilitate a broader understanding of the social and political tools that we already use to manage societal challenges. Further, the science-based solutions developed by applying a complex systems perspective to water management may also generalize to address many of the other intractable social and political challenges we face today, from addressing global poverty to mitigating climate change.

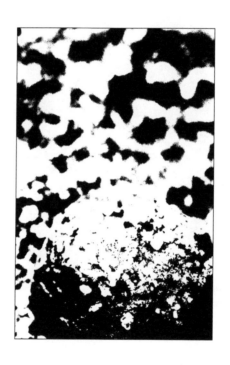

WHAT CAN MOTHER NATURE TEACH US ABOUT MANAGING FINANCIAL SYSTEMS?

*Simon Levin, Princeton University, and
Andrew Lo, Massachusetts Institute of Technology
Christian Science Monitor, August 22, 2016*

During a half-hour interval on May 6, 2010, stock prices for some of the largest companies in the world dropped precipitously, some to just pennies a share. Then, just as suddenly and inexplicably, shares recovered to their pre-crash prices.

This unprecedented event, burned into the memories of investors and regulators alike, is now known as the flash crash. Since that day, financial markets have seen flash crashes in US Treasury securities, foreign currencies, and exchange-traded funds (ETFs). Other puzzling, system-wide glitches are becoming more frequent as well.

Without a doubt, our financial systems are complex and often unpredictable, and when they swing out of control they remind us how much we still have to learn about how they work and how inadequate our traditional methods of controlling them are.

In all their complexity, though, financial markets don't hold a candle to the natural world, with its eight million-plus species—those we know of, not including the millions that have come and gone—interacting and evolving in the world's forests and oceans and in the microbiomes of our guts.

In the century and a half since Charles Darwin published *On the Origin of Species*, we still are stymied by the complexity of the biosphere, and, just as with our financial systems, our efforts to intervene have often led to confounding results.

Smokey Bear offers an example. For seven-plus decades, this popular icon has reminded us of the importance of preventing wildfires. But modern ecological practice recognizes that suppression

PHASE III: TERRAFORMERS

of all forest fires today simply sets up forests for larger and more destructive catastrophes tomorrow. In all likelihood, financial systems are no different; small catastrophes are probably essential in maintaining their ongoing health.

But how? How might biology inform our efforts to manage markets? How can we get beyond a metaphorical understanding of the ways markets and ecosystems are alike to explore, in practical terms, what our scientific theory offers our financial regulatory apparatus?

Earlier this year, during a Santa Fe Institute–sponsored meeting at the Keck Center of the National Academy of Sciences in Washington, DC, we pulled together experts from economists to ecologists to evolutionary biologists. We called the meeting "New Approaches to Financial Regulation." Our intent was to find common theoretical grounds with which to inform future financial regulatory approaches.

The Complex Systems Perspective

The biosphere and the "financiosphere" are both dazzling in their complexity, with striking similarities. Both are dynamic systems in which the selfish actions of countless individuals—whether they be cells or investors—lead to unpredictable consequences at the system level. In turn, these collective actions and consequences feed back to influence individual actions in endless cycles of adaptation and evolution.

This adaptive cycle is the essence of a complex system. It's also what makes complex systems difficult to understand, hard to predict, and tricky to manage.

Not surprisingly, in both the biosphere and financial markets, the resulting system-level emergent phenomena include unexpected crises and collapses, from population crashes to stock devaluations, from the desertification of lush landscapes to market failures, from the disappearance of species to the demise of industries.

But biological systems also exhibit remarkable resilience. By studying how evolution has made them more robust, might we develop new and wiser approaches to financial regulation? We think so.

Exploration and Exploitation

Life began on this planet nearly four billion years ago, and, despite frequent insults and challenges, we are still here (at least for the moment). We know that life's remarkable robustness, in large part, is dependent on variation; systems that suppress or lose their diversity are prone to collapse.

Through continuous innovation, via mutation and sexual recombination, for example, coupled with a seemingly simple filter called *natural selection*, which leads to the fittest innovations surviving to reproduce, life responds and adapts to changing environments and to itself. Charles Darwin, impressed by the "tangled bank" that emerged from these evolutionary dynamics, revolutionized our understanding of the world about us, and his insights are still with us.

But natural selection's apparent simplicity turns out to be deceptively complicated. Even the mechanisms of evolution, including those that generate innovation in the form of new variants, are subject to constant modification. Mutation rates (the rapidity at which genetic variants occur) are subject to selection pressures (influences that suppress a population's reproductive success). Even sexual reproduction itself has evolved to provide a greater variety of genetic material on which natural selection can act.

This interplay between exploration, by which new solutions are tested, and exploitation, by which the best solutions are multiplied and spread, is characteristic not only of evolution via natural

PHASE III: TERRAFORMERS

selection but also of the way people, companies, and other institutions must allocate their time and effort to survive and thrive in an economy—which is to say that business and markets are shaped by many of the same evolutionary processes that shape the natural world.

Evolving for the Unknown

Importantly, evolution is not about optimization in the abstract; it is about optimization relative to other genetic variants within and across species. While we are evolving, so too are our enemies (like the influenza virus) and our friends (including the microbiomes in our guts). To a large extent, evolution is about preparing for the unknown, because the scope of possible changes in our environments is so immense that we cannot hope to predict their form or timing.

We *can* predict, however, that during our lives, we will be assaulted by a variety of pathogens. Thus, vertebrates have evolved a contingency plan in the form of immune systems and barriers to invasion, such as skin and cell membranes. These systems combine early-warning indicators and generalized first lines of defense that buy time while we populate our immune repertoire with more specialized antibodies tuned to the specific threats. This is akin to circuit breakers in financial securities markets, which shut down trading when volatility is too high.

At the same time, mammals have evolved regulatory systems that help maintain the stability of our systems. Human heart rate and breathing, for example, are regulated by physiological processes that correct deviations from the norm in the timescales required for survival—kicking them into overdrive when we're being chased by a tiger, for example.

But when our physiological feedback loops are too weak or too slow, or too strong, pathologies arise.

Regulatory feedbacks that are "just right" help maintain a healthy human.

Similarly, when the timescale of financial innovation outstrips that of regulation—as in the case of high-frequency trading—there are likely to be unintended consequences. But regulatory responses that are too strong or poorly timed—like emergency price controls, short sales restrictions, bank holidays, and extreme capital constraints—can lead to greater panic and uncertainty among investors and consumers, ultimately causing even less desirable outcomes such as housing market crashes, rapid inflation, and recessions.

Regulatory feedbacks that are "just right" help maintain a healthy economy.

Self-Organized Robustness

These complex interrelationships underscore the importance of maintaining diversity in financial markets, in part by allowing enough exploration (that is, financial innovation) to produce the requisite diversity for a healthy system. But what is the right amount?

As mentioned earlier, evolution has dealt with the diversity problem in part by regulating evolutionary processes themselves: the rate at which mutations occur, and sexual recombination, which helps ensure a reassortment of the genes in a population and the production of new variants.

We tend to think of evolutionary change primarily in terms of natural selection based on the reproductive success of individuals with helpful traits. But all complex systems, including biological systems and business ecosystems, also exhibit self-organized patterns at scales larger than at the level of individuals. Such self-organization also "selects" by producing, from the interactions of individuals, emergent features that themselves either persist or wane. The self-organized systems that persist, and that we observe, tend to have properties that make them more robust.

PHASE III: TERRAFORMERS

Such self-organization does not always lead to robust systems, however; self-organized phenomena may also contain the seeds of system collapse, as we saw in the financial crisis of 2008–2009, when financial innovation and unprecedented connectedness, among other factors, combined to bring the system to the brink of failure.

Still, some features of biological systems might be helpful in designing self-organized financial systems for robustness.

Redundancy provides insurance against loss. The American chestnut largely disappeared from the forests of the northeastern United States, but other species filled its niche. In 2004, though, when Chiron, one of only two companies providing flu vaccines in the US, announced that its plants in Liverpool were contaminated, our house of cards was at real risk of collapse. We were too dependent on too few suppliers.

Modularity (the inverse of connectedness) isolates related elements, limiting systemic risk by reducing the potential for a local problem to spread globally. Quarantines and barriers restrict movement of infected individuals to help control the spread of a contagion. Such methods are used not only for human diseases but for livestock, as in the case of foot-and-mouth disease. Likewise, modularity in financial systems can help keep problems that emerge in one market or industry from spreading to others and pulling the whole system down.

In biology, breakdowns in size regulation, such as with gigantism, are considered unhealthy for biological organisms. Likewise, the unchecked growth of financial institutions can lead to banks that are "too big to fail," which, we now understand clearly, can threaten global financial stability.

Cues from Evolution

Any view of financial systems must recognize that they are ecosystems, linking agents, stocks, and flows. Just as an ecosystem ecologist is focused on the cycling of crucial elements like carbon, nitrogen, and phosphorus, so too might a "financial ecologist" focus on the sustainable cycling of crucial elements like capital, labor, and financial innovation.

As we refine and define the levers by which we attempt to manage tomorrow's economies, we must keep in mind that regulations that focus on specific parts of systems often miss the big picture. In the buildup to the 2008 crisis, for example, bank regulators naturally focused on the banking industry, neglecting the impact of the rapidly emerging shadow-banking system and its impact on financial stability.

Management of ecological systems in the past has often opted for narrowly derived or simplistic interventions, but the ensuing failures have led to calls for ecosystem approaches—in the management of fisheries and forests, for example. Similarly, our failures to predict and control financial ecologies should remind us that, if anything, the interconnectedness of global financial systems is ever greater, and a holistic approach is essential if we are to succeed.

As adaptive complex systems, natural and financial systems share deep likenesses. We should take cues from billions of years of evolution. Nature, and biology, offer solutions to a number of challenges of financial regulation, not to mention the regulation and control of many other systems crucial to well-functioning societies.

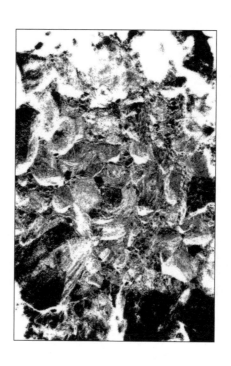

WHAT HAPPENS WHEN THE SYSTEMS WE RELY ON GO HAYWIRE?

John H. Miller, Carnegie Mellon University
Christian Science Monitor, September 19, 2016

Army ants can form a sixty-foot-wide, three-foot-deep front that moves through the forest like a bulldozer blade composed of millions of ants seeking prey. This behavior, essential to the colony's survival, is not directed by some central authority. The coordination emerges out of each ant's simple programmed responses to chemical signals.

At times, though, the colony's behavior can go awry. If, by chance, the marching ants happen to circle back upon themselves, they will follow one another in a circular mill, each ant dutifully obeying signals as they collectively march themselves to death.

The science of complex systems is the study of how local interactions can lead to global consequences. Army ants illustrate both the great promise and peril of systems exhibiting emergent behavior.

Well-functioning complex systems are essential to our collective survival. Consider the markets that provide us with everything from food to energy to entertainment to foreign trade to insurance and more. Markets are driven by the actions of countless individuals, each reacting to his or her whims, the weather, and the news of the day. Remarkably, just as with army ants, out of all our individual actions emerges a higher order, a set of prices that allows us to buy and sell whatever we may desire.

Each price contains a vast amount of information. The price of a gallon of gasoline, for example, incorporates everything from the weather in a far-off port to the stability of a foreign government. It tracks refinery availability in the Gulf of Mexico and emissions

PHASE III: TERRAFORMERS

regulations in California, while simultaneously taking into account the increased demand related to our summer vacation plans.

But where do prices come from?

Prices arise from the countless acts of many individuals attempting to trade in various goods. They are an emergent phenomenon.

Economists have recognized the magic of the market for hundreds of years. Even self-interested traders can, together, bring about the efficient allocation of the world's goods and services, and the use of markets has been a remarkable engine of our survival and success as a cooperative species. It's one of the best examples of the great promise of complex systems to improve our world.

There are well-known circumstances where markets fail, of course. Markets can be monopolized, leading to too-high prices. They can produce byproducts, such as the exploitation of finite natural resources, that harm individuals not involved in the immediate trade. Markets also tend to emphasize efficiency rather than equity. These failures can usually be corrected with appropriate public policy.

But there is also the potential for systemic failure, akin to the circular mill of army ants, tied to the complex nature of markets. What happens when emergence goes bad?

On May 6, 2010, around 2:30 p.m. Eastern time, key US equity indices plummeted for no apparent reason. The chaos in the indices initiated a tsunami that quickly began to wash over other markets. Stocks of formerly robust, mainstream companies began to trade at absurd prices. A share of Accenture, just minutes before trading at $40, could be had for a penny. A share of Apple shot from $250 to $100,000. The event, which lasted for about a half hour, is now known as the flash crash.

The proximate cause of the flash crash appears to have been a set of trades initiated by a money-management firm with an address in Shawnee Mission, Kansas. On that day, the firm wanted to sell a large block of shares.

Chapter 32: Systems Going Haywire

Key to selling a large block of shares is to slowly mete them out to the market so that other investors can't detect the glut and take advantage of the situation. Rather than assign a person to this task, the Kansas firm took a more modern approach and delegated the task to a computer program.

The devil is in the details of such programs, of course, and there was a devil indeed. The program, rather than relying on the current market price of the shares, kept selling as long as the firm's shares constituted only a small part of the overall market volume. Under usual conditions, such a strategy would ensure that the market was liquid and prices were reasonable.

But markets have become highly connected and computerized, so what happens in one market doesn't stay in one market. Information now flows quickly through the system, and computerized trading algorithms can be programmed to execute trades in the blink of an eye. (Actually, eye blinks take a poky 350 milliseconds, which is slow compared to today's algorithmic stock trades.)

If prices across markets are not perfectly aligned, an investor can buy the cheaper variant of a security in one market while simultaneously selling the more expensive version in another market, making a guaranteed profit.

These changes in trading technology have created a new kind of complex system, unforeseen even a decade ago. At the start of the flash crash, shares began to enter the market. Algorithms monitoring the various markets started rapid cycles of buying and selling to each other, resulting in a sudden increase in market volume.

This is where the fatal flaw in the Kansas firm's program became apparent: this increase in market volume caused the firm's algorithm to dump more shares onto the market, increasing the market volume even more, which in turn caused the algorithm to sell more. A positive feedback loop emerged and the original block of shares sold in less than twenty minutes. In the past, sales of similarly sized

PHASE III: TERRAFORMERS

lots took at least six hours. The complexity inherent in the market induced a human version of the circle of army ants.

The rapid influx of new shares into the market caused their prices to fall dramatically. This resulted in a misalignment of prices in other markets, and the chaos began to spread. The huge volumes of resulting transactions started to overwhelm the news feeds, which further exacerbated the crisis as both the machines and the humans in the loop could not make sense of the deluge of events.

The stock exchanges have built-in "circuit breakers" designed to halt trading when market conditions are such that the execution of additional trades would result in unnaturally large price swings. A five-second trading pause was imposed by one such mechanism thirteen minutes into the flash crash.

In a world where nanoseconds rule, five seconds is an eternity, and it appears that this action was enough to nudge the system toward more normal behavior thirty minutes into the event. The various stock exchanges involved in the flash crash recognized that the market conditions that prevailed during the event were not "fair and orderly" and that some of the prices that arose were "clearly erroneous." The trades that took place during the event were reversed over the next few days. The Securities and Exchange Commission implemented new circuit breakers into the markets, though their design was driven far more by intuition than by scientific objectivity.

Crisis averted. Lessons learned. And yet we might not have learned enough.

The 2010 flash crash was driven by ignorance and greed, but not malice. It is easy to imagine what could happen if malice and a bit more forethought were directed at disrupting our markets in similar ways.

We have created a complex, adaptive, and emergent financial system that we do not fully understand or know how to control. Each piece of this system makes sense: interconnecting markets allow arbitrage to keep prices in line, algorithmic trading ensures that there is always a willing trading partner, derivatives provide new means to hedge risk.

While each piece makes sense, the collective often does not. The study of complex systems suggests that knowing the behavior of each individual piece of a system does not give us insight into the behavior of the system as a whole. Implementing a circuit breaker in one market, for example, may resolve the immediate issue that market faces, yet shunt the problem to other markets.

We may well be at a stage where we cannot fully grasp the implications of the financial system we have built. Most of the time, out of this complex system emerges an order that allows us to thrive in a complicated world. Yet, relying on such a system also entails the potential of something going horribly wrong.

Our well-being now relies on the complex systems that bind our food supplies to our energy networks to our global climate to every institution in our society. The urbanization of our planet, the sustainability of our ecosystems, and the stability of our political systems are all entwined in complex, adaptive, interacting systems.

We get better as we go. We learn from every setback. But the feedback loops that will drive tomorrow's crises are already embedded in our systems, concealed by their complexities, and we don't know where they are or what might trigger them, much less how to control them.

Complexity is an aspect of our world that is amenable to scientific analysis, understanding, and perhaps even control, though we are only beginning to make progress on these fronts. We find ourselves in a race for knowledge and control of the complex world around us, a race that we must win if we are to thrive, and perhaps even survive, as a species.

WHEN AN ALLIANCE COMES WITH STRINGS ATTACHED

Paula L.W. Sabloff, SFI
Christian Science Monitor, November 4, 2016

I sat on the tarmac of the Chinggis Khaan International Airport outside of Ulaanbaatar, waiting for a plane to load and take off before mine. It contained 180 Mongolian soldiers on their way to Iraq to join George W. Bush's "coalition of the willing"—thirty-four nations committing troops to support the United States' 2003 invasion of Iraq. I was in Mongolia conducting anthropological research on Mongolians' perceptions of democracy.

Like several former Soviet countries, Mongolia wanted to show its solidarity with the United States. What better way to demonstrate its desire to make America its "third neighbor" than to support President Bush's war effort? (Mongolia is landlocked, sandwiched between the Russian Federation and China; "third neighbor" refers to Mongolia's interest in balancing its immediate neighbors with more Western-leaning nations.)

But did Mongolia want a true alliance with the United States, or was the relationship something else entirely?

It helps to define *alliance*. Although many people—including scholars—often call any cooperative relationship between two powers an alliance, not all alliances are the same. Nations or people in a true alliance treat each other as equals. They work together to meet a common goal—defeating or deterring an enemy, for example.

Allies make decisions together and they deal with the consequences of their decisions together. Once its goal is reached, an alliance may dissolve. The shared goal is the nucleus of the partnership.

Of the prominent US allies in the 2003 Iraq invasion, only the United Kingdom supported the war effort until 2009, when it

PHASE III: TERRAFORMERS

withdrew its troops. Australia withdrew its troops in 2003 but sent in more troops in 2005. France, Germany, and New Zealand, traditional US allies since the end of World War II, refused to participate in the war effort.

Partnerships of unequal powers are really patron–client relationships. Here, the relationship is based on a prior obligation rather than a partnership centered around a common goal.

The patron–client relationship always starts with a gift, and we know that there is no such thing as a free gift. If someone gives us a gift, then we are obliged to give something or do something in return. The gift giver becomes the patron of the gift receiver, who becomes the client.

In a patron–client relationship, when the patron has a goal such as winning a war or an election, he or she does not consult the client about goals or strategies. Rather, the patron may "call in his chips," requiring the client to help meet his goals by contributing in some way.

In the 2003 "coalition of the willing," Mongolia and most of the other participating nations were the clients of the United States. As recipients of US foreign aid, it was in their best interest to appear to be supporting the US in its war effort.

We often see patron–client relationships in politics. I saw this firsthand when I lived in Pennsylvania. There, state legislators had access to a slush fund that they could draw from to sponsor a firehouse or playground in their districts. The fund was called WAM, or "walking-around money." It enabled elected legislators to become patrons to their districts. Voters became their clients and presumably worked to get them reelected.

We see political patron–client political relationships operating at larger scales, too, such as in the current presidential election. We voters do not like to see our elected officials "in the pockets" of special interests, lobbies, and large donors. When we do, we assume

Chapter 33: When an Alliance Comes with Strings Attached

the politicians are clients in a patron–client relationship, and we assume their obligations will cloud their impartial judgment.

A good example is some people's concern about Hillary Clinton's ties to the Clinton Foundation and its acceptance of large donations while she is running for elected office. When Donald Trump claimed that he used his own money to finance his primary campaign, he was really declaring, to the chagrin of his party leadership, that he was not to be part of Washington's massive network of intertwining patron–client relationships.

Alliances and patron–client relationships are not new, of course. They've been a feature of human civilization for millennia, at least, and presumably these relationships have been critical in the formation of new political configurations, social institutions, and nation-states since the earliest state-like human societies.

The critical difference between alliances of equals and patron–client relationships became evident to me as I explored the roles of kings in newly emergent states.

The research was part of a Santa Fe Institute project, supported by a grant from the John Templeton Foundation, that is looking anew at the beginnings of state organization all over the world — from Mesopotamia and Egypt to China, Hawai'i, Mesoamerica, and South America.

In all eight places, territories were ruled by kings (or sometimes queens) who were believed to be inspired by divinity. Commoners in these societies farmed or made crafts, while a ruling elite managed their coordination and fought wars.

Surprisingly, such state-level organization arose independently in several societies around the world. Because these traditions had no contact with each other, we know they developed their state-formation patterns independently.

The Santa Fe Institute's project sought the common patterns in these seemingly separate emergent societies. My part of this research

PHASE III: TERRAFORMERS

was to figure out how rulers reduced their risk of losing wars. By comparing rulers' actions in traditional states, I found that both alliances and patron–client relations played important roles at least as far back as the beginnings of these eight states.

I also found that warfare was "built into" the political patterns of all the states. Rulers were expected to be warriors and military leaders who could expand or at least protect their existing territories. Sometimes, these rulers formed true alliances and fought together against a common enemy. In these cases they strategized together and divided up the spoils of war.

In 1428, the rulers of three Aztec towns—now part of Mexico City—formed a Triple Alliance. Together they conquered the Basin of Mexico. They divided it into three separate kingdoms but continued to work together to subdue smaller kingdoms, capture people for human sacrifice, demand tribute, and control trade routes. Once these goals were achieved, the alliance disintegrated; by the time the Spanish arrived in 1519, one of the partners, Tenochtitlan, had dominated its former allies and ruled the territory from the Pacific to the Atlantic.

But most rulers did not form true alliances, or partnerships of equals. Instead, those with greater military might turned the conquered rulers into clients. They often gave lavish gifts to secure the new patron–client relationship.

The greatest gift a conquering ruler could offer to secure a client's loyalty was his sister or daughter in marriage, and we have evidence that six of the eight early states in our study did this. These royal women carried the royal blood of the conqueror, thought to be superior to the royal blood of the conquered. By mingling superior blood with his own, the client ruler raised the prestige of his heirs, and so his line—his dynasty—could rise in rank in the region.

Of course, gifting a daughter or sister not only signaled the conquering ruler's superiority, it also had the advantage of turning the conquered into long-term clients, at least as long as the marriage

Chapter 33: When an Alliance Comes with Strings Attached

lasted. Trying to avoid client status, four of the societies (Egypt, Hawai'i, Aztec, and Inca) eventually opted for royal incest.

Sometimes client rulers wanted to break away from their patrons. They signaled this desire by abusing or trumping up charges against their royal wives so that they could legitimately execute them. Two client rulers of the Aztec refused to accept primary wives from their patron but married each other's daughters instead. In a true alliance, the brothers-in-law waged war against their superior.

Client rulers knew that they owed their patrons a great debt. When the time was right, patrons demanded military support—warriors, troops, and sometimes provisions—from their clients. Although we find patron–client relations in all the traditional states, we do not know when they first emerged. We do know they have persisted from one form of state organization to another.

Why would they persist, especially in situations of war? The simple answer is that they work. To win a war, it helps to have three things: more troops than the enemy, intelligence about the enemy's plans, and superior technology. Alliance-building and patron–client relations helped leaders amass more troops and often learn about the enemy's plans—and this is true whether a society is preindustrial, industrial, or postindustrial.

Politics in a democracy may be seen as a kind of war, although it is designed to be bloodless. There are winners and losers. Leaders build strategies for defeating the opposition. They ally themselves with equals, become patrons to constituents or lower-level candidates, and obligate themselves as clients to donors. But in the long run, the number of "troops"—the relative number of voters who actually vote—determines the winner.

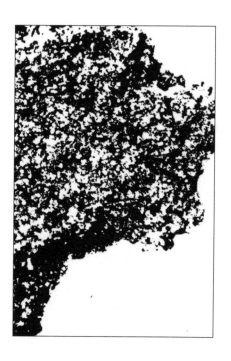

THANKSGIVING 2050: TO FEED THE WORLD WE HAVE TO STOP DESTROYING OUR SOIL

Molly Jahn, University of Wisconsin–Madison
Christian Science Monitor, November 23, 2016

At the height of the slave trade in 1785, an English divinity student, Thomas Clarkson, won a Latin essay contest considering the question, "Is it lawful to enslave the un-consenting?"

Few read it. Fewer took it seriously. But Clarkson, along with a small band of similarly inspired people, went to work, designing and executing a set of coordinated tactics to reveal the atrocities of legal slavery in the systems that brought sugar to British tables.

Wherever he went, Clarkson carried a wooden box filled with the slaver's tools—iron handcuffs, shackles, thumbscrews, branding irons, and instruments for forcing open slaves' jaws. Clarkson's moment of grace changed his course. Clarkson's box showed consumers the intolerable violence in their sugar bowls.

The violence that we do to our planet's soils, while by no means a crime comparable to the brutality of chattel slavery, is inseparably tied to our modern economic system, just as slavery was. And the mounting evidence of the violence we are doing to our soils is as obvious as the shackles in Thomas Clarkson's box.

The extractive farming methods that have been used since World War II to drive massive increases in agricultural yields and human population have brought our species and planet to a set of historic extremes—with unknown, but not unforeseeable, possibly devastating consequences for our food supply.

In May 2012, I lifted off from the Des Moines airport in a helicopter with a philanthropist and another scientist to take a look at America's soils. From above, the land looked tired, the beginning of the worst drought in twenty-five years. That afternoon, planters

PHASE III: TERRAFORMERS

scuttled hopefully back and forth in the haze, plumes of dust rising in the late-afternoon sun, symptoms of the damage we have inflicted on some of the richest soils on Earth.

Soil organic matter has dropped 30 percent to 50 percent since we began cultivating this ground. Compounded by erosion and agricultural practices that reduce soil life and damage soil structure, the injury to American soils is stunning and ongoing.

During these 150 years, our national anthem in agriculture has been "yield, yield, yield," the uncontested route to prosperity and abundance. But the harm we've done to our soils under this spell was plainly clear from the cockpit that afternoon in the rising trails of dust in the sky.

Always, the most dangerous lies are the lies we tell ourselves.

Despite millennia of traditional knowledge, and some major successes inspired by the Dust Bowl, it is estimated that nearly a third of the world's arable land has been eroded, lost at a rate of more than twenty million acres per year. The UN Convention to Combat Desertification estimates that more than 50 percent of land used for agriculture globally is "moderately or severely" degraded.

Injured soils lose resilience to drought and reduce agricultural productivity. When production fails, food systems more often go awry. When food systems go awry, people get frightened and desperate and angry. Chronic environmental stresses, somebody else's new dam, poor governance, and "subacute" events can and recently have added up to suffering, rebellion, violence, war, even the Syrian crisis.

By 2050, scientists estimate we will need to feed nine billion people with enormous implications for the world's resources. Already, geologists have graduated us from the Holocene, the comfy and stable geological period in which humankind evolved, to a new era called the Anthropocene. In the Anthropocene, humankind's actions shape the way our planet works, and agriculture is both our lifeline as a species and the dominant mode by which we care for our planet.

Yet, as British housewives once did, we are still earnestly ignoring the atrocities embedded in the food on our tables, and denying ourselves the possibility of much better. Our twentieth-century fragmented views, our inch-size glimpses of the elephant in the living room, do not reveal the dimensions of opportunities for radical innovations to better meet our needs—or a full understanding of the risks we face.

Further, our Whac-a-Mole responses to the range of symptoms we now detect, from degrading agricultural resources to epidemic obesity to mass species extinction, constrain our ability and our will to innovate. Our narrow views shackle us to incremental, stepwise change—and to the delusion that doing a little less bad and a little more good will be enough.

Always, the most dangerous lies are the lies we tell ourselves.

For all of human history, farmers have known that their soil is the living foundation of our species' future. For more than a century, scientists have formally confirmed these insights. Intensified and diversified crop and livestock rotations, "cover" crops, healthy microbial soil flora, no-till planting, and grazed, cropped systems are taking hold across America. These soil-care practices can slow the damage, even heal and regenerate soils while they boost and buffer agricultural productivity.

Innovations in the food and beverage sectors are also taking hold in the United States, which allow farmers to accelerate their commitments to improving soil health. Sustainability scorecards, wired-up farming systems, market differentiation, labeling innovations, and distributed sourcing are all schemes that allow farmers and

PHASE III: TERRAFORMERS

their supply chains to fulfill their commitments to best care for our soils. As these innovations take hold, we often find win-win-wins ... including improved profitability and resilience to extreme conditions.

But the questions remain: Are we moving in the right directions fast enough? Are we reducing the annual rate of soil loss and degradation? Can we objectively show we are actually starting to heal our soils? Is what we are doing enough to secure the agricultural future for our nation and our children?

The best science we have today signals the answer to these questions is *no*.

Thomas Clarkston started his lifelong journey toward abolition with a simple, now-famous observation: "if the contents of the essay are true, it is time some person should see these calamities to their end."

Today, a new science enables us to better see—and grasp—our challenges and opportunities. Complexity science investigates how relationships between parts give rise to the behaviors of a whole, how systems move, interact, form and reform.

Just as Clarkson's box and my view from the helicopter make obvious large-scale features of systems that we are unable or unwilling to see from the ground, complexity science can better illuminate the relationships between our individual and collective choices and their consequences. Cut free from disciplinary boundaries, complexity science systematically defines and reveals the elements of a system, how those elements may change or be changed, and how systems evolve.

These new views are like a "macroscope," bringing into sharp focus new ways of seeing what's already there. Complexity science, thus, can reveal strategies for systems change, the value of risk avoided, and open new paths forward. It can even reveal the power in a moment of grace—a phase change in the system sparked by an idea pushed into the system with forays of coordinated actions—with the potential for vast and amplifying benefits for humanity.

Advanced analytical approaches enable us to see how boundaries move, how feedback cycles work, how human and biophysical conditions interact. With new sight from billions of networked devices, and new science of the Earth system, human understanding of the violence, the tolerated-intolerables in our food systems is starting to change before our eyes.

The view from the helicopter cockpit that May afternoon was clear. Despite clear evidence to the contrary, the idea that American agriculture can be profitable *and* that our practices on the land must heal, even regenerate, our soils is still generally considered to be as preposterous as Clarkson's strong opinions about human bondage.

Last week, I flew into my home airport in Madison, Wisconsin. Looking down, I could see most of the harvested cornfields in my state famous for its progressive agriculture are "no-till" fields—fields where crop residues are left on the soil to "feed" the next crop and protect the soil from erosion. Other fields showed the dewy green of the winter wheat crop and cover crops that can hold the soil against the off-season winds and rains. Almost nowhere did I see the old-fashioned naked fields we used to leave open for erosion and exposure through the winter. Wavy "contours" of alfalfa, a perennial crop that restores fertility to the soil, and grassy waterways to limit erosion weave their way through the cornfields of western Wisconsin. Change is evident on the landscape.

But we know we can do better, and we are expecting a big crowd around the twenty-first-century table. The science says we have to do better. I say we must do better.

Armed with new scientific approaches to "see" our challenges, opportunities, and the benefits of the actions we take, armed with a wide range of social and technical innovations, bands of farmers, scientists, and many other partners in new cross-sector alliances are becoming empowered with new tools, broader views, and new inspiration to "see these calamities to their end."

HOW COMPLEXITY SCIENCE CAN HELP KEEP THE LIGHTS ON

Seth Blumsack, Penn State University
Christian Science Monitor, March 2, 2017

August 14, 2003, was a hot day in the northeastern United States, but not extremely so. Power lines carrying electricity to New York, Washington, DC, Toronto, and other major cities in North America were heavily taxed, but the loads were not unusual for an August afternoon.

On this particular Thursday, however, in rural Ohio, a power line happened to come into contact with a tree limb. This caused a power plant to go offline, then another. The shifting loads caused a Toronto-bound power surge, setting off a chain reaction that eventually brought some one hundred power plants down, affecting more than fifty-five million people in eight US states and Ontario and bringing the Northeast's commuter trains to a standstill.

In New York City, people walked down dozens of flights of stairs and then many miles more to their homes during the hottest part of a summer day. Air-conditioning systems failed, putting the elderly and the ill at risk. Phone service was interrupted as increased demand overloaded cell towers. Some municipal water systems lost pressure, prompting cities to advise their residents to boil their water before using it. Radio stations whose backup generators failed went off the air temporarily, frustrating the efforts of authorities to transmit emergency response instructions. Air transport and financial markets were disrupted. In some remote areas, power was out for a week.

An investigation led authorities to a bug in the software of an Ohio power company's control room that helped turn what should have been a manageable local blackout into a cascading regional failure. But the outage convinced even skeptics that the electrical grid was operating well beyond its design capabilities. Washington was swift to act, handing new responsibilities to the agencies that regulate operations and investment in the power grid.

Despite our best efforts, widespread power outages persist. Every several years now, a large portion of the US electrical grid collapses in similarly spectacular fashion, disrupting millions of lives. And each time, utility executives are called on the carpet. Billions of dollars are spent to harden the power grid. Laws are enacted to provide assurances that a disruptive power outage "never happens again."

And still, blackouts do happen, over and over. At the time, the northeastern blackout of 2003 was the second largest in world history, but since then, seven outages of even greater severity have occurred, with a 2012 blackout in India affecting one out of ten people on the planet.

Can disruptive blackouts be prevented? Does the ever-increasing complexity of our electrical supply system all but ensure more frequent and more catastrophic failures?

Unicycles and Spinning Plates

Electric power grids are marvelously complicated and intricate systems, comprising many millions of interconnected turbines, conductors, transmission lines, insulators, switches, and people. They tend to be enormous. The whole of the North American continent is served by just four or five regional grids.

The reasons for this complexity are perfectly sensible. For more than a century it has been cheaper to produce and distribute on a large scale, and cities and states have linked up their own utility grids with the growing network to increase redundancy (which added to their own systems' reliabilities) and to make trading in electrons possible. Bit by bit, the most intricate supply system ever created by humans took its form.

As a result, the behavior of our power grid is undeniably and irrevocably complex. The electricity that powers the glowing screen on which you could now be reading is the result of millions of interconnected devices working together in a highly synchronized way. Each of these elements behaves individually according to laws of physics that are easy to describe and predict.

> At the time, the northeastern blackout of 2003 was the second largest in world history, but since then, seven outages of even greater severity have occurred, with a 2012 blackout in India affecting one out of ten people on the planet.

But the system as a whole behaves in ways that are impossible to understand just by adding up the behaviors of these predictable parts. In other words, we know how the power grid works in theory. How it manages to work in practice is, even to trained professionals, often a mystery.

This complexity arises from a paradox: power grids are both inherently robust and inherently fragile. Operating a power grid is a bit like that old circus act of balancing spinning plates atop poles while riding a unicycle. Getting it all started is nearly impossible, but once you achieve an equilibrium, with all the plates spinning and the unicycle moving, maintaining that balance is somewhat easier—as long as you keep up the momentums of the various spinning parts.

Of course, this seemingly miraculous balance is delicate; the smallest upset can make the rider wobble. An overcorrection can invite disaster.

The engineers who designed the power grid know this and have built a tremendous amount of redundancy into the grid. On the grid, if a single plate falls and shatters, others are there to take its place. This redundancy makes our electricity supply remarkably reliable.

Self-Reinforcing Feedbacks

The behavior of the grid is, in its own way, an emergent one, meaning it arises from the interactions of many parts. But it's different from the collective behaviors of bee colonies or flocks of birds—in those, order seems to arise without the central control of a single decision maker. Members of the hive or flock, each following simple, programmed instructions, bring about self-organized, often surprising group-scale behaviors, like the undulating beauty of a murmuration of starlings.

> Under current policy for much of the US, the utility companies that would need to cooperate for such an arrangement to work are in direct economic competition with one another for electricity sales.

The power grid does have some ways of reinforcing its own stability. Because the elements in the power grid are so tightly connected, like gears in a machine, they exert a lot of force on one another. If one element becomes a little unstable, the others can compensate.

Chapter 35: How Complexity Science Can Help Keep the Lights On

But there are limits to the grid's ability to regulate itself. Like many other complex systems in which self-reinforcing anomalies can amplify themselves into powerful forces, the grid can destabilize itself to the point of destruction. If one element becomes unstable, it can trigger feedbacks that prompt the rest of the grid to overcompensate. The overcompensating equipment then becomes unstable, and the system destroys itself.

Thus, as with many complex systems created by humans, such as financial markets, the power grid needs people in the loop to control it and ensure that it operates in a stable manner. But also like financial markets, the modern grid confounds even its most highly skilled operators.

Power Grid Forensics is Hard

In 1997, a major power failure in the western US affected many millions of people. After months of study, engineers traced the problem to a piece of faulty equipment in southwestern Wyoming. Its failure overloaded two nearby pieces of equipment, so those failed as well, creating a kind of domino effect.

Then something strange happened. The failures of the first three pieces of equipment triggered the failure of a fourth—this one nearly 800 miles away on the Oregon–Idaho border. This failure was followed by a fifth at a substation in southwestern Montana, 400 miles to the east.

The chain of cascading failures played hopscotch throughout the western states, taking out nearly thirty pieces of critical power equipment until the slide was finally arrested somewhere in Nevada.

There are good reasons each of these pieces of equipment failed. But why the failures jumped around the way they did remains a mystery. Understanding the patterns of failures in cascading blackouts continues to be a challenge for power system operators and for scientists. No existing theory predicts or explains these patterns. The industry is learning more, but we are a long way from having the types of warning systems that are in place for earthquakes, tsunamis, and other disasters.

Are People the Problem or the Solution?

The northeast blackout of 2003 was, in part, instigated by the relatively mundane circumstance of a tree getting in the way of a sagging power line. As the blackout cascaded eastward, my own state of Pennsylvania was largely spared. We escaped the brunt of the blackout while surrounding states did not because the company operating my power grid made a spot decision to sever ties with the grids of neighboring states.

Which is to say that people and the decisions they make are a big contributor to the power grid's "robust fragility." We are the grid's fail-safe and its deepest weakness.

How? The ways people make investment and operations decisions about the power grid reflect two well-known dynamics in complex systems: preventing small localized problems can make future big problems worse, and competition can be more costly than cooperation.

We know from ecology that relentlessly preventing small problems can increase the risk of future big problems. For nearly a century, for example, the policy of the US Forest Service was to prevent all fires and, when that failed, to fight fires that did start early and aggressively. This was effective at preventing small and medium-sized fires and for protecting property. But it left behind fuel, which accumulated, turning future small fires into much bigger problems.

We have managed the power grid in much the same way. The small number of grids that serve North America is managed by a large number of utility companies, each of which is responsible for a small portion of the overall infrastructure.

Chapter 35: How Complexity Science Can Help Keep the Lights On

Those companies all make localized and quite rational decisions to reduce the frequency of outages within their individual operating footprints, which in practice means making investments to avoid relatively small blackouts in specific areas. This piecemeal approach, however, increases the danger of larger cascading failures with the potential to affect millions of people.

> We know from ecology that relentlessly preventing small problems can increase the risk of future big problems.

An illustration of the second effect can be seen in the collective behavior of animal herds. Caribou migrating across dangerous ice floes, for example, somehow can make good collective decisions if individuals are all striving toward the same goal—that is, if they cooperate. Similarly, computer simulations of the power grid by my colleagues Paul Hines and Sarosh Talukdar suggest that cooperative decision-making could keep blackouts from spreading if portions of the grid "island" themselves—if individual utilities effectively fall on their swords to save the grid as a whole. The process could be automated, as long as all the equipment on the grid was programmed with the same instructions in a sort of grand cooperative bargain.

But the reality of operating a power grid is very different. Under current policy for much of the US, the utility companies that would need to cooperate for such an arrangement to work are in direct economic competition with one another for electricity sales. And even if they weren't, individual utilities would still need to answer to state regulators for their own actions.

Some analyses have suggested that the 2003 blackout would not have spread so far had one utility company chosen to let the city of Cleveland go dark. In trying to keep the grid up to rescue Cleveland, that utility promulgated an unstable situation that spread to several other states and Canada.

And yet, imagine if those same utility executives had made a different decision, one that sacrificed Cleveland in the hopes of a greater good for the northeastern power grid. I would not want to be in the room when they had to explain to regulators in Ohio why they let some 20 percent of the state's population sit in the dark without air-conditioning in the dead of summer just to save New Yorkers from that fate.

An Integrated Human-Technological System

This collision of imperfect people, policy, and machines is why the power grid has fascinated complexity scientists for some time and has remained one of modern society's biggest challenges. Is the grid just a complicated system at massive scale whose modeling is made difficult by human interaction, or is it a technological manifestation of society's energy preferences?

Ultimately, as the 2003 blackout illustrates, it is both. The power grid is a collection of a large number of engineered devices, each of which has been programmed (like ants) to follow certain instructions, leading to highly coordinated collective behaviors. Those instructions, however, are the outcomes of social deliberations that range from the technocratic to the democratic.

So the grid is not just a massively complex infrastructure; it is one of the best examples we have of a truly integrated human-technological system.

This already-complex system is getting more complex by the day, as people decide to put solar panels on their rooftops, buy

Chapter 35: How Complexity Science Can Help Keep the Lights On

electric cars, or abandon the grid entirely. Are these preferences and decisions, which spread through communities much as diseases do, good or bad for the grid as a whole? The unfortunate answer at this point is that we don't know until we know—these are all (new) emergent behaviors whose ultimate outcomes are in play.

A great deal of progress has been made in recent years in using large amounts of data from the power grid to detect problems and, hopefully, solve them in real time. While these are stupendous capabilities, fundamentally they grapple with yesterday's grid. Our most urgent challenge is not to get better at describing what has happened but to understand whether it is possible to harness the complexity of the future power grid so it behaves more like an orderly bee colony and a lot less like a circus unicyclist.

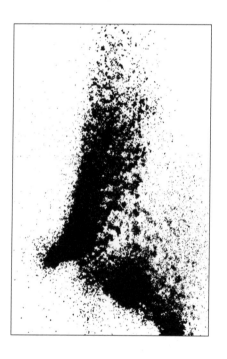

WHY PREDICTING THE FUTURE IS MORE THAN JUST HORSEPLAY

*Daniel B. Larremore, SFI, and
Aaron Clauset, University of Colorado Boulder
Christian Science Monitor, April 24, 2017*

Three years out of a PhD in physics in 1953, John Kelly, Jr. published a breakthrough paper about insider information in horse racing in an unlikely place: the Bell Labs Technical Journal. By the time it was in print, the paper's title had been scrubbed of its references to gambling—the AT&T executives didn't care for Bell Labs to be so directly associated with horse racing—but the content remained. Kelly had not just cracked the mathematics underlying a type of gambling, but he had also revealed deeper patterns about the nature of prediction.

When the odds posted by the track are different from the odds determined using insider information, Kelly's formula explains how to take those differences and place the best bets possible, mathematically speaking. The formula is powerful in its simplicity. It tells us to put money on every horse for which we have an informational or statistical edge, and then calculates exactly what fraction of our bankroll to bet on each horse, depending on the strength of that edge.

While this basic idea had long been known—the larger the difference in the track odds and the real odds, the bigger the opportunity for the gambler—Kelly quietly revolutionized the practice of prediction by writing down the optimal exchange rate between knowing something that others do not and the benefits of that knowledge.

Today, racetracks are less popular, but the principles remain the same. Asymmetries in the power to predict the future statistically

PHASE III: TERRAFORMERS

are the bread and butter of finance around the world, for example. But predictions underpin more than financial markets alone.

Prediction is the decoder ring of the modern world, touching everything from health care, car insurance, politics, and terrorism to sports, scientific discovery, and even the ride-hailing apps that are disrupting the taxi industry. In the age of bigger data and better algorithms, however, researchers are discovering straightforward systems that appear to be fundamentally unpredictable, as well as complicated systems whose behavior is surprisingly predictable.

Then again, this sort of thing is the norm when studying what researchers call complex systems—systems with many interacting elements whose collective behavior defies expectations based on their component parts. There may be simple patterns that organize seemingly chaotic events but complicated limits to prediction in rather simple systems. Finding the organizing patterns and challenging the limits to prediction are at the core of complex systems research.

Systems that involve people can be particularly surprising, because human agency would seem to make accurate prediction impossible. After all, if an equation predicts that a stock trader or public official will take a particular course of action, that person can simply take a different course, rendering the prediction immediately wrong.

And, while predicting what an individual might do is sometimes next to impossible, complex social systems can exhibit highly predictable behavior at large scales. For instance, no driver wants to get stuck in a traffic jam, but because of the choices each driver makes independently and the constraints of rush-hour travel, traffic jams emerge despite efforts to avoid them. Understanding the conditions under which they appear is fairly straightforward, even if no individual driver can predict which specific decisions will lead to a traffic jam.

Finding predictable patterns that emerge from the complicated interactions of many individual parts is the norm when studying

Chapter 36: Why Predicting the Future is More Than Just Horseplay

complex systems. Detecting these organizing patterns and outlining the limits of their predictability lies at the core of complexity science.

Deep Patterns in War and Violence

When an apple falls from a tree, everyone knows what happens next. We know from the application of the scientific method—that is, from observation, then explanation, then prediction, and finally verification—that gravity causes the apple to move toward the ground at a specific and constant rate of acceleration. Gravity and falling are so predictable that NASA engineers can hurl a satellite one hundred million miles across the solar system at Mars and still predict with an accuracy of a dozen feet where it will enter the atmosphere of the Red Planet.

> Prediction is the decoder ring of the modern world, touching everything from health care, car insurance, politics, and terrorism to sports, scientific discovery, and even the ride-hailing apps that are disrupting the taxi industry.

Human affairs are far messier. Take organized violence. Acts of terrorism can seem to occur at random places and times, and wars can erupt from causes as varied as internal political uprisings to territorial disputes. War and terrorism are archetypal chaos.

And yet, both wars and terrorism follow the same predictable mathematical pattern. In the early twentieth century, the English polymath Lewis Fry Richardson began looking at statistical

PHASE III: TERRAFORMERS

regularities in the sizes of wars, measured by the number of deaths they produce.

He discovered that wars follow nearly the same pattern as earthquakes. That is, just as the famous Gutenberg–Richter Law (on which the Richter scale is based) allows us to predict how many earthquakes of magnitude 3 or 4 or 6 will occur in California this year, Richardson's Law allows us to predict how many wars will occur over the next thirty years that kill ten thousand or fifty thousand or any other number of people.

Richardson's Law does not let us go beyond broad forecasts. It provides little help in predicting which countries will go to war, over what, and how large any particular war will be. Likewise, seismologists still struggle to predict precisely when or where any particular earthquake might occur or how large it might be.

In 1960, Richardson speculated that the statistical law that governs wars would hold for other types of violence, such as homicides or mass murders. Recent work suggests that he was not far off the mark. The same mathematical pattern as in the Gutenberg–Richter Law also appears in the sizes of terrorist attacks worldwide and may even hold for the mass shootings that have become disturbingly common in recent years.

Although these statistical regularities have improved our ability to estimate the broad brushstrokes of events, we still can't predict the precise details of the next mass murder.

Because we lack systematic data on the precise stresses and energy buildups in different parts of the Earth's crust, we cannot predict earthquakes. Similarly, the contingencies of human behavior make prediction that much harder within complex social systems, such as the ones that generate wars and terrorist attacks. We are far from having complete data. But, even if we did have it, we would not know what any particular person would do. We can predict patterns only at the global scale.

Chapter 36: Why Predicting the Future is More Than Just Horseplay

Certainty and Serendipity

Predicting the progress of science itself also runs aground on hard limits to accurate prediction. In 1964, Arno Penzias and Robert Woodrow Wilson discovered the cosmic microwave background, the noise left over from the early universe. They received a Nobel Prize in 1978 for their discovery that confirmed the big bang theory.

But Penzias and Wilson weren't even looking for the cosmic microwave background when they stumbled upon it. They were trying to detect radio waves bouncing off satellites that were carried to high altitudes by balloons. They kept getting a strange noise from their receiver. They tried to remove the noise by reorienting their antenna, by experimenting both day and night, and by clearing away a family of pigeons nesting in the antenna. But the noise remained. Only after eliminating all the alternative possibilities did they realize that, in fact, no radio source on Earth or even within our galaxy could explain their anomalous readings.

Penzias and Wilson had stumbled upon the echo of the big bang. Who could have predicted that? Alexander Fleming discovered penicillin in 1928 not through the deliberate and predictable processes of the conventional scientific method but by accident. CRISPR-Cas9, the wonder protein that enables scientists to edit a gene inside a living organism, was discovered by scientists studying an obscure aspect of certain bacteria.

You might think we scientists would be better at predicting discoveries. After all, scientists help choose which scientific projects receive support from taxpayers, which young researchers get hired to run their own labs, and which scientific papers survive peer review. Each of these choices is a kind of prediction that the scientific community depends on.

But the biggest discoveries are often the hardest to predict. We don't see them coming because they reorganize how we thought the world worked. Big discoveries are valuable precisely

PHASE III: TERRAFORMERS

because they are fresh and new, whereas predictions are always based on historical patterns.

Predicting that the future will be like the past, that accomplished scientists will continue doing good science, or that a hot area of research will continue to produce new ideas, is easy and natural. But it is also boring and shortsighted and therefore unlikely to hit upon truly unexpected ideas. Making those big leaps forward—the ones that change the way we understand the world around us—almost always requires a gamble with no guaranteed payout.

When Predictability is Boring

People like predictability. Predictability keeps us safe, ensuring that your car's airbag will deploy 99 percent of the time. Predictability allows us to anticipate and plan, and so homes in California must be built to withstand earthquakes. And predictability helps us avoid natural disasters, which is why we invest millions of dollars every year to operate weather satellites.

It may come as a surprise then, that in some systems, we actually seek to make things less predictable.

Professional basketball, it turns out, is one of these systems. Although millions of fans may feel otherwise, physicists and mathematicians have shown that the ups and downs of lead sizes over the course of a game are highly unpredictable. So unpredictable, in fact, that the outcome of most games is only slightly less predictable than guessing whether there will be more heads than tails when flipping one hundred coins.

Of course, you can improve your predictions about how a game will evolve if you know something about the teams, with separate calculations for offensive and defensive strengths, star players, coaching acumen, injuries, and the like. Even so, flipping coins to model how a game evolves will do a good job of predicting the outcomes of games over a whole season.

Chapter 36: Why Predicting the Future is More Than Just Horseplay

This dramatic unpredictability is puzzling. After all, professional athletes spend enormous energy honing their skills, and teams win or lose depending on how well their players play. To make this less puzzling, let's consider the spectators.

The fans love exciting games. Huge blowouts are no fun. Team sports are best when their ups and downs, and ultimately their outcomes, are as random as possible. A great game is one in which the teams are evenly matched and they battle valiantly We love it when an underdog manages an upset victory in the nick of time. In other words, we crave limited predictability in our sports.

In the early 1950s, basketball had become a boring game to watch. When one team opened up a lead, the game turned into keep-away, allowing the leading team to effectively freeze the score. There was no randomness, no level playing field, no promise of an exciting finish. Once a team gained a good lead, spectators might as well have headed for the car.

> People like predictability. Predictability keeps us safe, ensuring that your car's airbag will deploy 99 percent of the time. It may come as a surprise then, that in some systems, we actually seek to make things less predictable.

That's why Danny Biasone, owner of the Syracuse Nationals, advocated for, and eventually won, the introduction of the 24-second shot clock in the 1954–55 season. The shot clock, which requires a team to make a shot within twenty-four seconds of gaining possession of the ball, unfreezes the score and infuses a greater degree of randomness

PHASE III: TERRAFORMERS

into every game from beginning to end. In other words, in sports, we crave unpredictability so much that when a game becomes too predictable, we happily change the rules to make it more uncertain.

Unpredictability is an essential ingredient in any form of entertainment. A horror movie isn't fun without surprising frights, the best jokes always turn on an unexpected element, and love stories are appealing because it seems impossible that the two people will ever get together, yet they do so against all odds.

In team sports and entertainment, then, the limits to prediction are at the core of our interest. Unpredictability keeps us on the edges of our seats and can delight us or break our hearts. The same limits are the source of the gambler's love of sports, and the dream that one person's insider information can somehow be translated into a statistical edge at the bookie's desk.

The future of prediction

Ben Mezrich's book *Bringing Down the House* tells an exciting but fictionalized story of the MIT Blackjack Team. But in the 1990s, the real MIT Blackjack Team did go to Las Vegas. Armed with statistics, they turned their edge into cash and walked away with fortunes.

They weren't the first. The threads of their ideas reach back through history. In the late 1970s, a team from Santa Cruz built miniaturized computers, which they hid in their shoes and used to predict the clattering fate of the roulette wheel. Like the MIT students who would come after them, their predictions weren't perfect, but they knew that any statistical edge could be turned into winnings.

Earlier still, Ed Thorpe's victories over blackjack in the mid-1960s were the product of carefully exploiting the differences between good and bad prediction. He wrote the book on counting cards in blackjack, and started a hedge fund. By 1988, Thorpe's

Chapter 36: Why Predicting the Future is More Than Just Horseplay

personal investments had grown at an annualized rate of 20 percent per year for over twenty-eight years.

Thorpe's story begins earlier, too. He built the world's first wearable computer with Claude Shannon, a scientist at Bell Labs who fathered the age of computers with information theory. In 1961, two decades before the Santa Cruz students, Thorpe and Shannon built enough of a statistical advantage to beat Nevada's roulette wheels. Claude Shannon worked at Bell Labs with none other than John Kelly, Jr.

Kelly never gambled himself, but his formulas taught each of the players who followed how to convert prediction into earnings. In 1953, he quantified the fundamental value of prediction by equating an information edge with earnings and used the horses at the racetrack to illustrate his points. Although most of us have never heard of Kelly, today we use his ideas when making predictions about every part of our complex society.

When the stakes of prediction are high and the unexpected occurs, it's tempting to throw the baby out with the bathwater and fire the statisticians for the surprises they told us were unlikely. We would, of course, be unwise to do so.

In spite of its limits, the future of prediction has never looked brighter. Those who walk away from statistical predictions are leaving money on the table. Eventually, in the long run, they'll be on the losing side of people willing to read John Kelly's work.

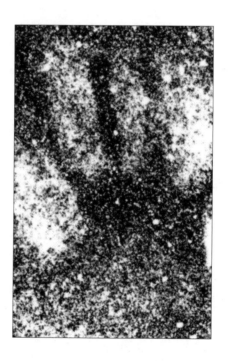

EMERGENT ENGINEERING: REFRAMING THE GRAND CHALLENGE FOR THE 21ST CENTURY

David C. Krakauer, SFI
Santa Fe Institute, 2019

How can human society have reached the moon, harnessed the unobservable mechanics of the atom, continued to build computers that become exponentially faster and cheaper each year, and yet have operated so poorly in establishing stable economies, reducing the incidence of conflict and disease, and discovering and manufacturing effective biomedical drugs? It is certainly not through lack of interest, resources, effort, and intelligence.

The war on cancer, the pursuit of greater economic equality and financial stability, the creation of online safety and security, and the invention of new nontoxic and effective pharmacological drugs have absorbed astronomical sums of money into both research and development—and yet in so many cases they have foundered and failed through the misapplication of previously highly successful ideas of engineering and design to complex systems.

There is an urgent need for novel concepts directed at achieving an evolutionary and emergent engineering, and it is our contention that they are likely to come from the domains of biological and social life—not from the deterministic world of designed mechanical artifacts.

A History of Success and Failure: The Siren Song of the Grand Challenge

There are a handful of technology projects of such sheer audacity and scale that they have become bywords for human ambition, ingenuity, and impact. Included among these projects should certainly be numbered the Apollo Program, the Manhattan Project, CERN

PHASE III: TERRAFORMERS

and the LHC, the Human Genome Project, the Panama Canal, and the Great Wall of China. Comparably impressive in scale and cost is the continued application of some version of Moore's law to integrated circuit design.

> Over the last several decades, new discoveries and theories have emerged to study complex systems—networks of adaptive agents—that promise to be better suited to addressing economics, disease, cybersecurity, and medical treatment than erstwhile approaches founded on very successful (where appropriate) classical engineering axioms.

The Apollo project, launched by President Kennedy in 1961 as an "urgent national need," expended $25.4 billion (around $165 billion, adjusted) and employed over four hundred thousand workers between 1959 and 1970. For the sake of comparison, the vastly less edifying financial bailout of 2008 (an effort to correct the technical deficiencies of designed market instruments) cost just short of $750 billion.

The Manhattan project was launched to produce nuclear weapons in response to the global threat from Nazi Germany and its allies. From 1939 to 1946, around 130 thousand people were employed at a cost of $2 billion ($22 billion, adjusted), spread across thirty sites, to produce a bomb. The so-called war on cancer (borrowing the martial metaphor to instill a disciplined sense of urgency and determination, and to justify huge expenditures) has been waged since it was first declared by Nixon in 1971. Research costs have been on the order of $100 billion, with the greatest gains in longevity coming through modified behavior and improved screening, whereas biomedical engineering has produced rather modest increases (a few percentage points) in life extension, according to Decennial Life Table data.

Moore's law describes the doubling of transistor density every two years from 1970 until the near present. This exponential trend has been accompanied by similar trends in reduced cost and increased memory capacity. The economic burden maintaining these trends has been considerable, including the cost of fabrication—a modern fabrication plant costs around $2 billion to build—and the percentage of revenue consumed by research and development increased by 40 percent over the decade from 1999 to 2009. Compare this to the deliciously impudent Eroom's law (Moore's law reversed), which describes how drug discovery since the 1980s has become more expensive, slowed down delivery of drugs to those in need, and produced ineffective and doubtful remedies—and this despite the giddy hyperbole surrounding drug design, machine learning, and high-throughput technologies.

Over the last several decades, new discoveries and theories have emerged to study complex systems—networks of adaptive agents—that promise to be better suited to addressing economics, disease, cybersecurity, and medical treatment than erstwhile approaches founded on very successful (where appropriate) classical engineering axioms.

The Classical Engineering Axioms

1. Design according to well-understood scientific principles that hold for all components in isolation and in aggregate.

PHASE III: TERRAFORMERS

2. Design systems with as near as possible fault-free components to very high levels of combined precision.

3. Minimize error and accept only the smallest system failure rates by eliminating uncertainty and reducing degrees of freedom of components.

4. Design systems into linear ranges of operation where collective dynamics are predictable and controllable.

5. Reduce noise and adaptability of components to prevent unexpected emergent behaviors.

To calibrate the difficulty of applying this kind of framework to complex problems, consider an appropriately complementary list of properties found in most complex systems of the kind that govern the societies in which we live, from the immune system to the nervous system, as well as ecosystems, economies, and political institutions.

The Properties of Complex Systems

1. We have few general design principles for adaptive components (cells, organisms, nations) in isolation or in the aggregate where new unforeseen properties emerge.

2. Components typically have high failure rates in all tasks and accomplish their objectives through statistical averaging and approximation across multiple scales and levels.

3. There is significant uncertainty and lack of information at both the component and aggregate level, and components have large—and often poorly understood—repertoires of behavior.

4. Most evolved complex systems operate in nonlinear and often near-critical regimes (close to thresholds and tipping points).

5. Adaptability of components is the rule—not the exception—and learning and adaptation are ongoing and irrepressible.

Nearly every assumption listed in the engineering axioms is violated by complex systems. So, what leads us to believe that we can use the insights of classical engineering to predict and control these systems? The recent historical record makes the case rather clear: we have not succeeded in this approach.

This does not mean that there is not enormous value in engineering design, or that we should give up. There is an alternative approach that respects in its own axioms the properties of complex systems and builds on the insights of classical engineering—we call this approach emergent engineering.

The Objectives of an Emergent Engineering

1. Seek to modify the reward or selective context in which semi-autonomous agents operate and design toward better incentives.
2. Accept significant component error rates and focus on mechanisms that can average and aggregate these effects to acceptable levels in the collective output.
3. Design with an eye towards distributions of outcomes and not towards deltas (single optimal outcomes), pursuing average properties throughout.
4. Develop mechanisms for controlling nonlinear dynamics and predicting and influencing critical transitions.
5. Harness adaptation to allow for continued exploration and exploitation rather than coercing systems into single states that require endless iterations of costly novel production.

This is obviously a very ambitious list, and yet there are both established and nascent areas of research, several coming from within engineering, where ideas with these characteristics have

PHASE III: TERRAFORMERS

been—and are being—developed. These include nonlinear control theory, evolutionary dynamics, genetic algorithms and programming, pattern formation, agent-based modeling, mechanism design, collective computation, programmable and adaptive matter, and distributed token systems such as the blockchain.

There are three very successful areas where these and related ideas have been applied by recruiting and not obstructing the properties of complex adaptive systems: vaccine development, auction design, and neural prosthetics.

Co-opting Complexity for Emergent Engineering

A vaccine is a chemical agent that provokes a sustained reaction from the immune system of a host directed at pathogens with features similar to the vaccine. Vaccines achieve emergent engineering by exploiting the noisy, distributed, nonlinear adaptive properties of immune systems. Vaccines are effective precisely because the contribution to immunity engineered into a vaccine is negligible in comparison to the host mechanisms that it engenders and that it exploits.

An auction is a mechanism for meeting the market needs of heterogeneous buyers and sellers without prior knowledge of pricing. The auction mechanism exploits uncertainty and the adaptive and strategic self-interests of buyers and sellers to achieve a desired distribution. Auction mechanisms are effective because they are simple and shift the burden of information processing and decision making to populations of agents and away from overly complicated, over-regulated, centralized valuation mechanisms.

Neuroprosthetics are electrical brain implants that are able to compensate for the loss of essential cognitive functions to include perception and movement. The effectiveness of neuroprosthetics relies on the brain's ability to learn the inputs and outputs of the implant. An effective implant is one that exploits the noisy adaptive nature of neural tissue such that the restoration of function is largely a matter of ongoing brain rewiring and recoding and more modestly reliant on the prewiring/pre-encoding of the prosthetic.

Each of these examples illustrates engineered solutions working cooperatively with a complex system in order to predict and control a behavior rather than shoehorn complexity into regimes where the classical engineering axioms hold sway. Success depends on a deep understanding and application of the properties of complex systems in a hybrid agent–artifact setting.

Next Steps

Engineering-inspired paradigms from genomics to synthetic biology and circuit-based neuroscience, upwards into market regulation and cybersecurity, need to continue to evolve by merging in part with the adaptive sciences to embrace complex reality and communicate the practices and implications of this new approach to the critical challenges of the modern world. The disruption of the disciplines and the growth of respect fostered across the spurious borders of the institutionalized sciences—departments and schools—would be a welcome and necessary corollary.

Perhaps the greatest "phase transition" in our thinking that such an approach could engender is the maturation in our willingness to live with relatively high levels of uncertainty in the domains of complex phenomena—and thus give up on ideas like complete "cures," the elimination of "risk," the design of perfect "stability," and achieving total "security." We replace these ideals of a deterministic age with an understanding of the ever-evolving nature of adaptive processes, seeking to discover new methods for the specification of incentives, rewards, constraints, and communication, together capable of moving outcomes into a space of desirable, albeit never optimal, performance.

DAVID C. KRAKAUER is the president and William H. Miller Professor of Complex Systems at the Santa Fe Institute. A graduate of the University of London, where he went on to earn degrees in biology, and computer science, Krakauer received his D.Phil. in evolutionary theory from the University of Oxford in 1995. He remained at Oxford as a postdoctoral research fellow, and two years later was named a Wellcome Research Fellow in mathematical biology and lecturer at Pembroke College. In 1999, he accepted an appointment to the Institute for Advanced Study in Princeton and served as visiting professor of evolution at Princeton University. He moved on to the Santa Fe Institute as a professor three years later and was made faculty chair in 2009.

Krakauer also served as director of the Wisconsin Institute for Discovery and the co-director of the Center for Complexity and Collective Computation, and was a professor of genetics at the University of Wisconsin, Madison. He has been a visiting fellow at the Genomics Frontiers Institute at the University of Pennsylvania and a Sage Fellow at the Sage Center for the Study of the Mind at the University of Santa Barbara.

In 2012, Krakauer was included in the *Wired* Magazine Smart List as one of 50 people "who will change the World." He is currently writing a book on intelligence, natural and artificial, and our historical lack thereof.

SFI PRESS BOARD OF ADVISORS

Nihat Ay
Professor, Max Planck Institute
for Mathematics in the Sciences;
SFI Resident Faculty

Sam Bowles
Professor, University of Siena;
SFI Resident Faculty

Jennifer Dunne
SFI Resident Faculty;
SFI Vice President for Science

Andrew Feldstein
CEO, Co-CIO & Partner, Blue Mountain
Capital Management; SFI Trustee

Jessica Flack
SFI Resident Faculty

Mirta Galesic
SFI Resident Faculty

Murray Gell-Mann
SFI Distinguished Fellow & Life Trustee

Chris Kempes
SFI Resident Faculty

Michael Lachmann
SFI Resident Faculty

Manfred Laublicher
President's Professor, Arizona State
University; SFI External Faculty

Michael Mauboussin
Managing Director, Global Financial
Strategies; SFI Trustee &
Chairman of the Board

Cormac McCarthy
Author; SFI Trustee

Ian McKinnon
Founding Partner, Sandia
Holdings LLC; SFI Trustee

John Miller
Professor, Carnegie Mellon University;
SFI Science Steering Committee Chair

William H. Miller
Chairman & CEO, Miller
Value Partners; SFI Trustee
& Chairman Emeritus

Cristopher Moore
SFI Resident Faculty

Mercedes Pascual
Professor, University of Chicago;
SFI Science Board Co-Chair

Sidney Redner
SFI Resident Faculty

Daniel Rockmore
Professor, Dartmouth College;
SFI External Faculty

Jim Rutt
JPR Ventures; SFI Trustee

Daniel Schrag
Professor, Harvard University;
SFI Science Board Co-Chair

Geoffrey West
SFI Resident Faculty; SFI Distinguished
Professor & Past President

David Wolpert
SFI Resident Faculty

EDITORIAL

David C. Krakauer
Publisher/Editor-in-Chief

Tim Taylor
Aide-de-Camp

Laura Egley Taylor
Manager, SFI Press

Sienna Latham
Editorial Coordinator

Katherine Mast
SFI Press Associate

THE SANTA FE INSTITUTE PRESS

The SFI Press endeavors to communicate the best of complexity science and to capture a sense of the diversity, range, breadth, excitement, and ambition of research at the Santa Fe Institute. To provide a distillation of discussions, debates, and meetings across a range of influential and nascent topics.

To change the way we think.

SEMINAR SERIES

New findings emerging from the Institute's ongoing working groups and research projects, for an audience of interdisciplinary scholars and practitioners.

ARCHIVE SERIES

Fresh editions of classic texts from the complexity canon, spanning the Institute's thirty years of advancing the field.

COMPASS SERIES

Provoking, exploratory volumes aiming to build complexity literacy in the humanities, industry, and the curious public.

For forthcoming titles, inquiries, or news about the Press, contact us at
SFIPRESS@SANTAFE.EDU

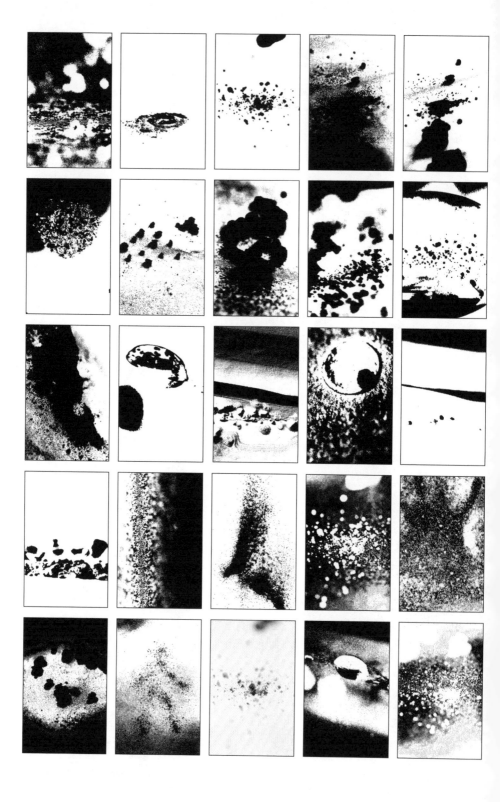

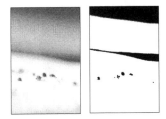

IMAGES IN THIS VOLUME

The images in this volume originated as photographs of sand pulled from the grounds of Santa Fe Institute's Cowan Campus. Each photo was then manipulated digitally and iteratively, using Adobe Photoshop.

Photographic illustrations by Laura Egley Taylor and Katherine Mast.

COLOPHON

The body copy for this book was set in EB Garamond, a typeface designed by Georg Duffner after the Egenolff-Berner type specimen of 1592. Headings are in Kurier, a typeface created by Janusz M. Nowacki, based on typefaces by the Polish typographer Małgorzata Budyta. The SFI Press complexity glyphs used throughout this book were designed by Brian Crandall Williams.

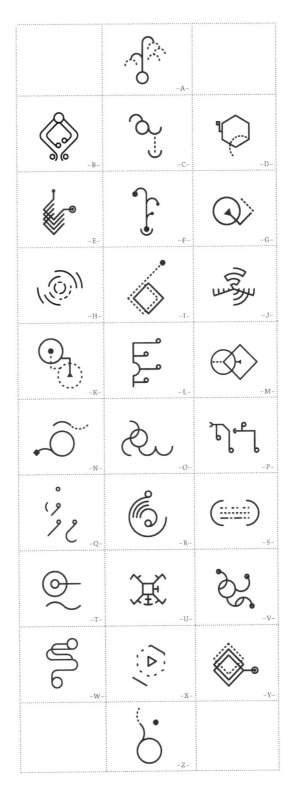

ABOUT THE SANTA FE INSTITUTE

The Santa Fe Institute is the world headquarters for complexity science, operated as an independent, nonprofit research and education center located in Santa Fe, New Mexico. Our researchers endeavor to understand and unify the underlying, shared patterns in complex physical, biological, social, cultural, technological, and even possible astrobiological worlds. Our global research network of scholars spans borders, departments, and disciplines, bringing together curious minds steeped in rigorous logical, mathematical, and computational reasoning. As we reveal the unseen mechanisms and processes that shape these evolving worlds, we seek to use this understanding to promote the well-being of humankind and of life on Earth.

COMPASS SERIES

Made in the USA
Columbia, SC
27 July 2021